BRIAN BLAIS

STATISTICAL INFERENCE FOR EVERYONE

LIFE'S MOST IMPORTANT QUESTIONS ARE, FOR THE MOST PART, NOTHING BUT PROBABILITY PROBLEMS.

PIERRE-SIMON LAPLACE

STATISTICAL THINKING WILL ONE DAY BE AS NECESSARY FOR EFFICIENT CITIZENSHIP AS THE ABILITY TO READ AND WRITE.

H.G.WELLS

STATISTICS ARE THE HEART OF DEMOCRACY.

SIMEON STRUNSKY

BRIAN BLAIS

STATISTICAL INFERENCE FOR EVERYONE

SAVE THE BROCCOLI PUBLISHING

PUBLISHED BY SAVE THE BROCCOLI PUBLISHING

TYPESET WITH TUFTE-LATEX

First printing, August 2014. Last Compiled August 26, 2014.

Dedicated to all of the wonderful people at Bryant University in the Library, Writing Center, and the Center for Teaching and Learning who have been exceedingly supportive of me through the entire process of writing.

Contents

List of Figures

List of Examples

List of Tables

Proposal

I would like to propose a new introductory statistical inference textbook, which I believe takes a fresh look at a course that fits into nearly every quantitative major at universities.

Initial Motivation

My motivation for this project stems from my dissatisfaction with traditional approaches to the topic, and my belief that there is a better way. A first semester statistics course is generally divided into the following four parts:

I Basic Statistical Concepts

- Basic statistical concepts including population, parameter, sample, and statistic
- Types of data (ordinal, time-series, etc...), and sampling methodology
- Organizing the data visually or graphically - including histograms, pie graphs, box plots, and stem-and-leaf plots
- Statistical computations including mean, median, mode, standard deviation, and percentiles

II Probability

- Properties of unions, intersections, conditional probability, independence and mutual exclusivity
- Permutations and combinations
- Discrete distributions
- Continuous distributions
- Normal distribution

III One-sample Statistics

- Confidence intervals

- Sampling distributions
- Computations involving the normal distribution, t-distribution, and binomial distribution (for proportions)
- Hypothesis testing

IV Two-sample Statistics

- Two sample problems - expanding topics from Part III to two variables

Obviously, there is some variability to these topics, but as one can see from most introductory statistics textbooks, there is a consistent approach. My main concerns about the traditional approach can be summarized as follows:

1. Part II (probability) generally covers at least one quarter of the material in an introductory statistics course. There is a shift from data collection and analysis (Part I) to probability theory. Subsequently, Part III shifts back to a data centered approach and only a small portion of Part II generally applies in Part III. This disconnect between Parts I, II, and III, impedes the learning process. It seems to the students as if the parts are related somehow, but the connection is rarely made. The students are then left with a feeling that the course concerns two completely unrelated topics: probability and statistics.

2. The normal distribution is covered repetitively throughout many chapters of most introductory statistics books. The coverage is included in sections such as: empirical bell-shaped curve (Part I), normal distribution as a type of continuous distribution (Part II), sampling distributions (Part III), interval estimation (Part III), hypothesis testing (Part III), and two population testing (Part IV). There is redundant focus on the normal and t-distributions. These topics are closely related, but not handled cohesively. More importantly, there is little or no discussion of the assumptions of the normal model or how to tell what constitutes "close enough" to normal. In addition, there is generally equal consideration given to the rare practical situation in which the standard deviation is known (and knowing this does not generally alter the result much at all).

3. After the concept of a "statistic" is covered, there are many chapters which repeat essentially the same problem multiple times, from only slightly different perspectives. This gives the student a feeling that these are all very different problems, despite the appearances, and leads the student to approach solving problems

like a "cookbook": just find the right recipe for the right problem. The fundamental understanding of statistical inference is undermined by this approach.

It is my view that the traditional approach detracts from student understanding, with its "cookbook" perspective, disjointed coverage of probability, and the almost exclusionary focus on the normal distribution.

A New Approach

In the field of statistical inference, there are two primary schools of thought. Each has its proponents, but it is generally accepted that on all problems covered in an introductory course, that both approaches are valid and lead to the same numerical values when applied to actual problems. Only one of these approaches is covered in a traditional course, which denies the students access to an entire field of statistical inference. The traditional approach, also called the frequentist or orthodox perspective, leads almost directly to problem (1) above. The other approach, also called Probability Theory as Logic[1], derives all statistical inference from probability theory directly. It is this approach that I hope to expose students to in an introductory course.

[1] E. T. Jaynes. *Probability Theory: The Logic of Science*. Cambridge University Press, Cambridge, 2003. Edited by G. Larry Bretthorst

The probability theory approach to statistical inference has several benefits:

1 All of the same problems as handled traditionally can be handled with this perspective, yielding *exactly the same answers*[2].

[2] One reason why "Probability theory as Logic" concepts are covered only in advanced courses is the misperception that they are applicable only to more advanced problems, and not applicable to problems normally found in an introductory class. The fact that this misperception exists is a strong argument for a book like this one, to dispel this misperception and to communicate both to students and instructors alike the value of a this approach to basic problems.

2 Statistical inference is theoretically grounded in probability theory, which, although admittedly beyond an introductory course, avoids the "cookbook" approach, where different problems need different methods, that students take away from the traditional textbooks. Here all problems use the *same* method, derived from probability theory.

3 The reasoning process using the probability theory perspective is more intuitive than the orthodox perspective, especially when dealing with hypothesis testing.

   For example, every statistics instructor faces the challenge of getting students to interpret p-values properly, and the logic behind setting up null-hypotheses. They have to combat the students' initial intuition that the p-value represents the "probability that the null is true," and many students never really obtain the proper understanding. I have even heard instructors use it this way.

In the Probability Theory as Logic perspective, this same calculated value is interpreted *exactly like the students' initial intuition*! Thus, testing hypotheses, estimating parameters, and determining uncertainties are far more direct and intuitive using this approach than the traditional approach.

What I Am Proposing

This text can help solve the challenges described above, and more. By focusing on models and data, as opposed to populations and samples, this text can more cohesively bridge the topics described in Parts I, II, and III above. Probability will be introduced as a natural part of solving problems, as opposed to its standalone treatment traditionally done in today's texts.

In this text, I will use the Probability Theory as Logic approach applied to the same problems that are traditionally covered. This viewpoint can greatly enhance our understanding of statistics and can handle topics such as confidence intervals and hypothesis testing in a very intuitive manner. Statistical inference covered in this way also addresses real-life questions that are not addressed by traditional statistical methods.[3]

[3] One of the reasons why this approach is usually covered only in more advanced courses is the difficulty of the mathematics generally associated with it. Orthodox statistics makes heavy use of sampling, which is deemed more intuitive than probability distributions. It is my intention to start with low-dimensional cases, building to distributions, and to augment all concepts with numerical exercises.

Finally, this will be a problem oriented textbook. It is imperative that the problems are cohesive with the pedagogy. I will also plan to use technology, where appropriate, to further student learning and make the textbook more interactive.

At the level targeted for this book, there is only one textbook that I know of that covers inference from the perspective proposed here, and that is Donald Berry's book *Statistics: A Bayesian Perspective*, 1996. It is my intention to modernize the approach, and include some topics that are not covered, specifically from the physical sciences and business.

1 Introduction to Probability

Life's most important questions are, for the most part, nothing but probability problems. - Laplace

In 1968 a jury found defendant Malcolm Ricardo Collins and his wife defendant Janet Louise Collins guilty of second degree robbery[1]. The decision hinged on the testimony of bystanders, which stated that the perpetrators had been "black male, with a beard and moustache, and a caucasian female with blonde hair tied in a ponytail," and that they escaped in a "yellow motor car." A mathematician testified that the odds *against* this couple being innocent were one in *twelve million*, and this was enough for the jury to convict. Later, in an appeal, the California Supreme Court reversed the decision primarily because of lack of evidence, and faulty inference.

[1] J. Sullivan. People v. Collins , 68 cal.2d 319, 1968. URL http://scocal.stanford.edu/opinion/people-v-collins-22583

In another case, Sally Clark was convicted in 1999 of the murder of her two young sons[2]. Again, the testimony hinged on a statistical argument - the chances of one baby dying in their bed 1 in 8500, so therefore the chances of *two* of them dying in the same way is the square of this, or 1 in 73 million. Several years later, and a public statement from the Royal Statistical Society highlighting the erroneous logic, Sally Clark was released - although she never overcame the resulting damage to her life that the conviction had caused.

[2] Lord Justice Kay. R vs Sally Clark, April 2003. URL http://www.bailii.org/ew/cases/EWCA/Crim/2003/1020.html

We will cover these cases in more detail later, and why the inference was faulty, but I introduce the stories here for two reasons. First, is to point out that there are cases in which proper statistical inference can be a life and death matter. Second, it is to highlight the fact that such inference can run counter to one's intuition. Part of the purpose of this book is to retrain your intuitions and your habits of intuition to avoid such failures.

We have to make decisions nearly every second of our lives, and those decisions are based on our state of knowledge. Unfortunately, we are never 100% sure of *any* information in our lives, so we are constantly forced to make decisions in the face of uncertainty. In many cases our common sense is enough to make sophisticated decisions, taking into account the uncertain nature of the situation. However, there are many times where our common sense is not enough to quantitatively resolve the level of uncertainty, and make valid inferences. It is in these cases that statistical inference is most useful.

Statistical inference refers to a field of study where we try to infer unknown properties of the world, given our observed data, in the face of uncertainty. It is a mathematical framework to quantify what our common sense says in many situations, but allows us to exceed our common sense in cases where common sense is not enough. Ignorance of proper statistical inference leads to poor decisions and wasted money. As with ignorance in any other field, ignorance of statistical inference can also allow others to manipulate you, convincing you of the truth of something that is false.

For example, in 1978 a Russian satellite deviated from its orbit and became increasingly erratic, and was going to crash into the Earth.[3] This sort of event occurs from time to time, even including a recent crash of a US spy satellite in 2008.[4] There was a local news broadcast about the impending Russian satellite crash which said something like, "the scientists had studied the trajectory of the satellite, and determined that there was only a 25% chance of it striking land, and even a much smaller chance striking a populated area." The report was clearly designed to calm the public, and convince them that the scientists had a good handle on the situation. Unfortunately, given a little thought, one realizes that the Earth's surface consists of about 25% land and 75% water, so *if you knew nothing about the trajectory of the satellite*, you would simply state that it had a 25% chance of striking land. Instead of communicating knowledge of the situation, the news broadcast communicated (to those who knew basic statistical inference) that either the scientists were in *complete ignorance* of the trajectory or the reporter had misinterpreted a casual statement about probabilities and didn't realize what was implied. Either way, the intent of the message and the content of the message (to those who understood basic probability) were in direct conflict.

[3] L Heaps. *Operation morning light*. Paddington, S.l, 1978. ISBN 0709203233

[4] James Oberg. U.S. satellite shoot-down: The inside story. *IEEE Spectrum*, 2008

1.1 Models and Data

There are two main aspects of statistical inference: description of data and model analysis. In the description of data, one attempts to summarize a set of data with a smaller set of numbers. Grades in the classroom are summarized by the average, votes in a state are summarized by a percentage, etc... This smaller description of the data is useful for both practical and theoretical reasons. It is more expedient to communicate a small set of numbers than the entire data set, and it is almost always the case that the detailed properties of a set of data are not relevant to the questions that you are asking.

A model refers to a mathematical structure which is used to approximate the underlying causes of the data, and unify seemingly unrelated problems. One may have a (mathematical) model for a coin flip which ignores all of the details of the flip, the bounce, and the catch and summarizes the possible results by a single number: the chance that the coin will come up heads. You may then use that same model to describe the voting behavior of citizens during a presidential election, or to describe the radioactive decay of particles in a physics experiment. The mathematics is identical, but the interpretation of the components of the model will be different depending on the problem. Models *simplify*, by summarizing data with a small set of causes, and they are used for *inference*, allowing one to predict the outcome of subsequent events.

The goal of statistical inference is then to take data, and update our knowledge about various possible models that can describe the data. This often means deciding which of several models is the most likely. It can also entail the refinement of a single model, given the new data. All of these activities are closely related to (and perhaps identical to) the methods in science. What we are trying to do is make the best inferences from the data, improve our inferences as new data come in, and plan what data would be the most useful to improve our inferences. In a nutshell, the approach is:

$$\text{Initial Inference} + \text{New Data} \rightarrow \text{Improved Inference}$$

In order to deal with a wide variety of problems, we require a minimal amount of mathematical structure and notation, which we introduce in this chapter.

1.2 *What is Probability?*

Probability theory is nothing but common sense reduced to calculation. - Laplace

When you think about probability, the first things that might come to mind are coin flips ("there's a 50-50 chance of landing heads"), weather reports ("there's a 20% chance of rain today"), and political polls ("the incumbent candidate is leading the challenger 53% to 47%"). When we speak about probability, we speak about a percentage chance (0%-100%) for something to happen, although we often write the percentage as a decimal number, between 0 and 1. If the probability of an event is 0 then it is the same as saying that *you are certain that the event will never happen*. If the probability is 1 then *you are certain that it* will *happen*. Life is full of uncertainty, so we assign a number somewhere between 0 and 1 to describe our state of knowledge of the certainty of an event. The probability that you will get struck by lightning sometime in your life is $p = 0.0002$, or 1 out of 5000. Statistical inference is simply the inference in the presence of uncertainty. We try to make the best decisions we can, given incomplete information.

In this book, our approach is to determine, for each problem, what degree of confidence we have in all of the possible outcomes. The approach of statistical inference covered in this book is about the procedure of most rationally assigning various degrees of confidence (which we call *probability*) to the possible outcomes of some process using all the objectively available data.

One can think of probability as a mathematical short-hand for the common sense statements we make in the presence of uncertainty. This short-hand, however, becomes a very powerful tool when our common sense is not up to the task of handling the complexity of a problem. Thus, we will start with examples that will perhaps seem simple and obvious, and move to examples where it would be a challenge for you to determine the answer without the power of statistical inference.

Let's walk through a simple set of examples to establish the notation, and some of the basic mathematical properties of probabilities.

Card Game

A simple game can be used to explore all of the facets of probability. We use a standard set of cards (Figure 1.1) as the starting point, and use this system to set up the intuition, as well as the mathematical notation and structure for approaching probability problems.

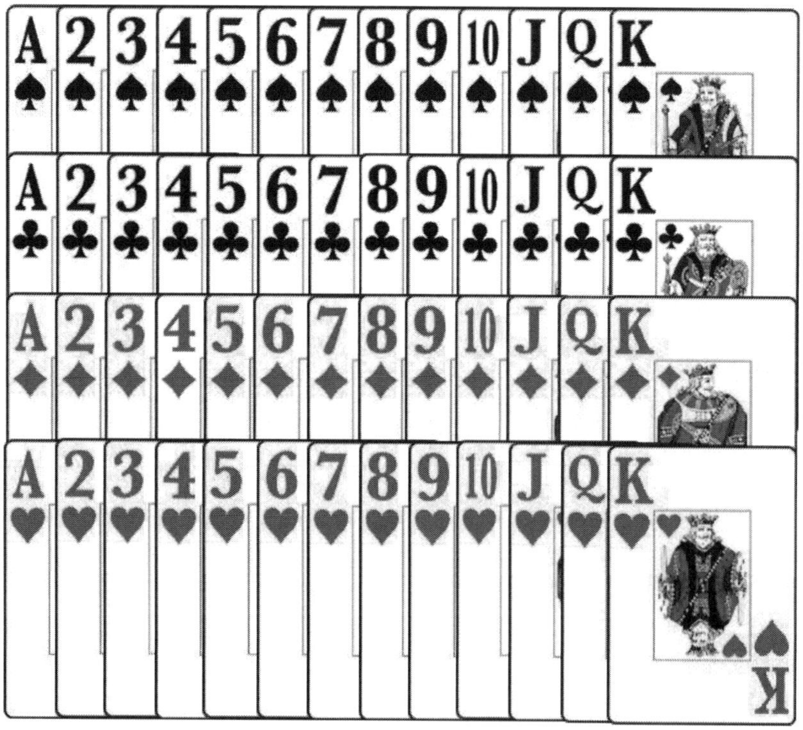

Figure 1.1: Standard 52-card deck. 13 cards of each suit, labeled Spades, Clubs, Diamonds, Hearts.

We start with what I simply call the *simple card game*[5], which goes

[5] In this description of the game, we do not reshuffle after each draw. The differences between this non-reshuffled version and the one with reshuffling will be explored later, but will only change some small details in the outcomes.

like:

$$\text{simple card game} \equiv \begin{cases} \text{From a standard initially shuffled} \\ \text{deck, we draw one card, note what} \\ \text{card it is and set it aside. We then} \\ \text{draw another card, note what card} \\ \text{it is and set it aside. Continue until} \\ \text{there are no more cards, noting each} \\ \text{one along the way.} \end{cases} \quad (1.1)$$

There are certain principles that guide us in developing the mathematical structure of probability. We start with some common sense notions, written in English, and then write them as general principles. These principles, then, constrain our mathematics so that we can apply the ideas *quantitatively*.

When asked "what is the probability of drawing a red on the first draw?" you would generally say 50-50, or 50%, or equivalently written as a probability, $P(R_1) = 0.5$. The reason for this is that we are completely ignorant of the initial conditions of the deck (i.e. where each card is located in the deck after the initial shuffling). Given this level of (or lack of) knowledge, we could swap the colors of the two suits and we would have an equivalent state of knowledge - the problem would be identical. We will keep coming back to this concept, but in general:

Principle of Knowledge and Probability Equivalent states of knowledge must yield equivalent probability assignments.

Principle of Knowledge and Probability Equivalent states of knowledge must yield equivalent probability assignments.

Because of this principle, we are led to the conclusion that

$$P(R_1) = P(B_1)$$

where R_1 represents the statement "a red on the first draw" and B_1 represents "a black on the first draw." Because these are the only two options, and they are mutually exclusive, then they must add up to 1. Thus we have

$$P(R_1) = 1 - P(B_1)$$

which leads directly to our original assignment

$$P(R_1) = P(B_1) = 0.5$$

Mutually Exclusive If I have a list of *mutually exclusive* events, then that means that only one of them could possibly be true. Example events include flipping heads or tails with a coins, rolling a 1, 2, 3, 4, 5 or 6 on dice, or drawing a red or black card from a deck of

Mutually Exclusive If I have a list of *mutually exclusive* events, then that means that only one of them could possibly be true. Examples includes the heads and tails outcomes of coins, or the values of standard 6-sided dice. In terms of probability, this means that, for events A and B, $P(A \textbf{ and } B) = 0$.

cards. In terms of probability, this means that, for events A and B, $P(A \textbf{ and } B) = 0$.

Non Mutually Exclusive If I have a list of events that are *not mutually exclusive*, then it is possible for two or more to be true. Examples include weather with rain and clouds or holding the high and the low card in a poker game.

Non Mutually Exclusive If I have a list of events that are *not mutually exclusive*, then it is possible for two or more to be true. Examples include weather with rain and clouds or holding the high and the low card in a poker game.

Now, this was a long-winded way to get to the answer we knew from the start, but that is how it must begin. We start working things out where our common sense is strong, so that we know we are proceeding correctly. We can then, confidently, apply the tools in places where our common sense is not strong.

In summary, with no more information than that there are two mutually exclusive possibilities, we assign equal probability to both. If there are only two colors of cards in equal amounts, red and black, then the probability of drawing a red is $P(R_1) = 0.5$ and the probability for a black is the same, $P(B_1) = 0.5$.

Other Observations

If instead of just the color, we were interested in the suit (hearts, diamonds, spades, and clubs), then there would be four equal and mutually exclusive possibilities. We have a certain number of possibilities, and our state of knowledge is exactly the same if we simply swap around the labels on the cards. If we're interested in the specific card, not just the suit, the logic is the same. Thus, we have

$$P(\spadesuit) = P(\clubsuit) = P(\blacklozenge) = P(\heartsuit)$$

and for drawing one specific card from the deck,

$$P(A\spadesuit) = P(2\spadesuit) = P(3\spadesuit) = \cdots = P(K\heartsuit)$$

Further, they all must add up to 1, so we get for suits

$$P(\spadesuit) + P(\clubsuit) + P(\blacklozenge) + P(\heartsuit) = 1$$

and for the specific card from the deck,

$$\underbrace{P(A\spadesuit) + P(2\spadesuit) + P(3\spadesuit) + \cdots + P(K\heartsuit)}_{\text{52 cards}} = 1$$

Putting it together, we get for the suits

$$P(\spadesuit) = P(\clubsuit) = P(\blacklozenge) = P(\heartsuit) = \frac{1}{4}$$

and for the specific card

$$P(A\spadesuit) = P(2\spadesuit) = P(3\spadesuit) = \cdots = P(K\heartsuit) = \frac{1}{52}$$

Probabilities for Mutually Exclusive Events In general, for mutually exclusive events, we have

$$P(A) = \frac{\text{(number of cases favorable to A)}}{\text{(total number of equally possible cases)}} \quad (1.2)$$

Probabilities for Mutually Exclusive Events

$$P(A) = \frac{\text{(number of cases favorable to A)}}{\text{(total number of equally possible cases)}}$$

1.3 *Conditional Probability*

It is important to understand that probability reflects our state of knowledge about the system. As our knowledge changes, so do our probability assignments. As we gain more information, we change our probability assignments. Two people observing the same system, but with *different* information about the system, will give *different* probability assignments. All we need to make sure probability theory matches our common sense is for two people with the same state of knowledge, or the same information, to yield identical probability assignments.

Because our information about a system is so important in assigning probabilities, we introduce a way of writing it mathematically that we will use for the rest of the book. It will be good for the reader to get used to reading the mathematical short-hand in English in order to gain an understanding for what it means.

Probability Notation

In math, we choose to abbreviate long sentences in English, in order to use the economy of symbols. In this book we choose a middle-ground between mathematical succinctness and the ease of understanding English. We start with the simple card game (Equation 1.1)

We then define a new symbol, |, which should be read as "given." When there is information given we call this probability *conditional* on that information. When we write the following:

$$P(\text{red on first draw}|\text{simple card game}) \quad (1.3)$$

or

$$P(R_1|\text{simple card game}) \quad (1.4)$$

this is short for

"The probability of drawing a red on the first draw, **given that** *we have a standard initially shuffled deck and we follow the procedure where we draw one card, note what color it is and set it aside and continue drawing, noting, and setting aside until there are no more cards."*

One can easily see that the mathematical notation is far more efficient. It is important to be able to read the notation, because it describes what we know and what we want to know.

Conditional Probability When information is given, and expressed on the right-hand side of the | sign, we say that the probability is *conditional.* $P(\text{I'm going to get wet today}|\text{raining outside})$ is an assessment of how likely it is that I will get wet *given*, or *conditional on*, the fact that it is raining outside. Clearly this number will be different if it was conditional on the fact that it is sunny outside - different states of knowledge yield different probability assignments.

When we put a comma (",") on the right side then we read this as "and we know that." For example, when we write the following:

$$P(\text{red on second draw}|\text{simple card game,red on first draw}) \quad (1.5)$$

or

$$P(R_2|\text{simple card game}, R_1) \quad (1.6)$$

this is short for

"The probability of drawing a red on the second draw, **given that** *we have a standard initially shuffled deck and we follow the procedure where we draw one card, note what color it is and set it aside and continue drawing, noting, and setting aside until there are no more cards* **and we know** *that we drew a red on the first draw."*

Conditional Probability When information is given, and expressed on the right-hand side of the | sign, we say that the probability is *conditional.* $P(\text{I'm going to get wet today}|\text{raining outside})$ is an assessment of how likely it is that I will get wet *given*, or *conditional on*, the fact that it is raining outside. Clearly this number will be different if it was conditional on the fact that it is sunny outside.

Causation. Imagine we have a 2-card game: a small deck with one red card and one black card, and I draw a red card first. Clearly this makes the probability of drawing red as the second card equal to zero - it can't happen. We're tempted to interpret

$$P(R_2|R_1, \text{2-card game}) = 0$$

to mean that *because we drew a red on the first draw, this* **causes** *the impossibility of drawing the red on the second* - there is only 1 red card after all, and drawing it seems to **cause** the impossibility of drawing red in the future. However, consider the following:

$$P(R_1|R_2, \text{2-card game}) = 0$$

which is, if we *knew* that the second card we drew was red, then it makes it impossible to have drawn a red card as the first card. This is just as true as the previous case, however, you can't interpret this as *causation* - the second draw didn't *cause* the first draw.

Instead, *probability statements are statements of logic, not causation.* One can use probabilities to describe causation (i.e. $P(\text{rain}|\text{clouds})$), but the statements of probability have no time component - later draws from the deck of cards act exactly the same as earlier ones.

1.4 *Rules of Probability*

From the rule for mutually exclusive events (Equation 1.2), we assign the following probabilities for the *first draw* from this deck[6]:

- $P(10) = \frac{4}{52}$
- $P(\heartsuit) = \frac{13}{52} = \frac{1}{4}$
- $P(10\heartsuit) = \frac{1}{52}$
- $P(\text{face card}) = \frac{12}{52}$
- $P(\text{number card}) = \frac{40}{52}$

[6] A face card is defined to be a Jack, Queen, or King. A number card is defined to be Ace (i.e. 1) through 10.

It turns out that mathematically, the rules for *fractions of things* and of *probabilities* are the same. Thus, to gain an understanding for the rules of probability, we will calculate fractions (which are more immediately intuitive), and then summarize the same rule for probabilities.

Negation Rule

In this section I'll use the letter F for fraction, and we can determine the values simply by counting. The fraction of cards which are hearts (♥) is

$$F(\heartsuit) = \frac{13}{52} = \frac{1}{4}$$

The fraction of cards which are *not* hearts (i.e. the 3 other suits) is:

$$F(\text{not } \heartsuit) = \frac{13 \times 3}{52} = \frac{3}{4}$$

These numbers add up to one: $F(\heartsuit) + F(\text{not } \heartsuit) = 1$. We can do this with more complex statements.

$$\begin{aligned} F\left(\text{first card is a face card}\right) &= \frac{12}{52} \\ F\left(\text{first card is not a face card}\right) &= \frac{40}{52} \\ F\left(\text{first card is a face card}\right) + F\left(\text{first card is not a face card}\right) &= 1 \end{aligned}$$

EXAMPLE 1.1 *What is the fraction of the first card as a jack given that we know that the first card is a face card?*

We can also apply the negation rule to conditional statements, like "the first card is a jack given that we know that the first card is a face card." Notice that there are 12 cards that are face cards, so we restrict our counts to those.

$$\begin{aligned} F\left(\text{jack}|\text{face card}\right) &= \frac{4}{12} = 1/3 \\ F\left(\text{not a jack}|\text{face card}\right) &= \frac{8}{12} = 2/3 \\ F\left(\text{jack}|\text{face card}\right) + F\left(\text{not a jack}|\text{face card}\right) &= 1 \end{aligned}$$

and they add up to one.

Either-or fallacy. The negation rule, should not be taken to imply that everything is "black and white," or "there are only two sides to every story." It really is just a statement of logic, should be carefully considered and has some limitations. For example, the following is true,

$$P\left(\text{object is black}\right) + P\left(\text{object is not black}\right) = 1$$

However, this does not mean the same thing as

$$P\left(\text{object is black}\right) + P\left(\text{object is white}\right) \neq 1$$

"Not black" is not the equivalent of "white." It could be red, or gray, or some other color. A common logical fallacy sometimes referred to as the "either-or fallacy" or the "fallacy of the excluded middle," turns on this point. Some examples of these fallacies are:

- If we don't reduce public spending, our economy will collapse.
- You're either with us or you're a terrorist.
- Either modern medicine can explain how Ms. X was cured, or it is a miracle.

Negation Rule Given any information, we have

$$P(\textit{statement}|\textit{information}) + P(\textbf{not } \textit{statement}|\textit{information}) = 1$$

or

$$P(A|B) + P(\textbf{not } A|B) = 1 \tag{1.7}$$

Negation Rule

$$P(A|B) + P(\textbf{not } A|B) = 1$$

Product Rule

The product rule comes from looking at the combination of events: event A *and* event B. As before, we'll work on the numbers from the fractions of the card game.

EXAMPLE 1.2 *What is the fraction of cards that are Jacks and a heart?*

This is clearly $F(\text{J}\heartsuit) = 1/52$, but we can look at it a different way that is equivalent. We note that the Jacks constitute 4/52 of the cards, and that *of those* 4, only one quarter of them are hearts (one card out of the four cards). So, we can arrive at the fraction of J♥ by taking one quarter of the fraction of jacks. So what we have is

$$F(\text{jack } \textbf{and } \heartsuit) = F(\heartsuit|\text{jack}) \times F(\text{jack}) = \frac{1}{4} \times \frac{4}{52} = \frac{1}{52}$$

One can equivalently reason from the suit first: the hearts constitute 13/52 of the cards, and that *of those* 13, the Jacks constitute 1/13 of the cards. So, we can arrive at the fraction of J♥ by taking one thirteenth of the fraction of ♥. Again, we have

$$F(\text{jack } \textbf{and } \heartsuit) = F(\text{jack}|\heartsuit) \times F(\heartsuit) = \frac{1}{13} \times \frac{13}{52} = \frac{1}{52}$$

In general we have

Product Rule

$$P(A \text{ and } B) = P(A|B)P(B) = P(B|A)P(A) \quad (1.8)$$

Product Rule

$$\begin{aligned} P(A \text{ and } B) &= P(A|B)P(B) \\ &= P(B|A)P(A) \end{aligned}$$

EXAMPLE 1.3 *What is the probability of drawing two Kings in a row?*

This is the same as

$$P(K_2 \text{ and } K_1)$$

From the product rule (Equation 1.8) we have

$$P(K_2 \text{ and } K_1) = P(K_2|K_1)P(K_1)$$

The second part is straight forward: $P(K_1) = 4/52$. The first part is asking the probability of drawing a second king, knowing that we have drawn a king on the first draw. Now, there are only 51 cards remaining when we do the second draw, and only 3 kings. Thus, we have $P(K_2|K_1) = 3/51$ and finally

$$\begin{aligned} P(K_2 \text{ and } K_1) &= P(K_2|K_1)P(K_1) \\ &= \frac{3}{51} \times \frac{4}{52} = \frac{1}{221} \end{aligned}$$

Independence

As a specific case of the product rule, we can change the rule of the card games such that we reshuffle the deck after each draw. In this way, the result of one draw gives you no information about other draws. In this case, the events are considered *independent*.

Independent Events Two events, A and B, are said to be independent if knowledge of one gives you no information on the other. Mathematically, this means

$$P(A|B) = P(A)$$

and

$$P(B|A) = P(B)$$

Independent Events Two events, A and B, are said to be independent if knowledge of one gives you no information on the other. Mathematically, this means

$$P(A|B) = P(A)$$

and

$$P(B|A) = P(B)$$

In this case, the product rule reduces to the simplified rule for independent events: the product of the individual event probabilities.

Joint Probabilities for Independent Events

$$P\left(A \textbf{ and } B\right) = P(A) \times P(B) \tag{1.9}$$

Joint Probabilities for Independent Events

$$P\left(A \textbf{ and } B\right) = P(A) \times P(B)$$

We have already seen an example of this, when we looked at drawing the Jack of Hearts: drawing a heart gives you no information about whether it is a jack, and vice versa. Thus,

$$P\left(\heartsuit|\text{jack}\right) = P\left(\heartsuit\right)$$

EXAMPLE 1.4 *What is the probability of flipping two heads in a row?*

The probability of getting "heads" on any given coin flip is $P(H) = 0.5$. The probability of flipping two heads in a row is then simply $P(H_1) \times P(H_2) = 0.5 \times 0.5 = 0.25$, because the second flip is independent of the first. If it wasn't, then you'd have to determine how the knowledge of the first flip influences our knowledge of the second flip, which is written as $P(H_2|H_1)$ and the full product rule (Equation 1.8) would need to be used.

Conjunction

One of the consequences of combinations of events is that the probability of two events happening, A **and** B, has to be less than (or possibly equal to) the probability of just one of them, say A, happening. The mathematical fact is seen by looking at the magnitude of the terms in the product rule

$$P(A \text{ and } B) = \underbrace{P(B|A)}_{\text{less than or equal to } 1} \times P(A) \leq P(A)$$

In other words, coincidences are less likely than either event happening individually. We intuitively know this, when we make comments like "Wow! What is the chances of that?" referring to, say, someone winning the lottery and then getting struck by a car the next day. Sometimes, however, it seems as if one's intuition does not match the conclusions of the rules of probability. One such case is called the *conjunction fallacy*.

In an interesting experiment, Tversky and Kahneman[Tversky and Kahneman, 1983] gave the following survey:

> Linda is 31 years old, single, outspoken, and very bright. She majored in philosophy. As a student, she was deeply concerned with issues of discrimination and social justice, and also participated in anti-nuclear demonstrations.
>
> Which is more probable?
>
> 1 Linda is a bank teller.
>
> 2 Linda is a bank teller and is active in the feminist movement.

85% chose option 2.[Tversky and Kahneman, 1974] This, they attributed, to the conjunction fallacy - mistaking the conjunction of two events as more probable than a single event. They went further and did a survey of medical internists with the following

> Which is more likely: the victim of an embolism (clot in the lung) will experience partial paralysis or that the victim will experience both partial paralysis and shortness of breath?

and again, 91 percent of the doctors chose that the clot was less likely to cause the rare paralysis rather than to cause the combination of the rare paralysis and the common shortness of breath.

Even when correct, the consequence for conjunctions can be misused, or at least misidentified. Returning to our example of someone winning the lottery and then getting struck by a car the next day, rare events *occur frequently*, as long as you have enough events. There are millions of people each day playing the lottery, and millions getting struck by cars each day. We will explore this problem later in Section 2.2, but one immediate consequence is that winning the lottery and getting struck by a car the next day probably happens *somewhere* fairly regularly.

Combinations of Events and the English language I believe that the issue of the conjunction fallacy is more subtle than this. In English, if I were to say "Do you want steak for dinner, or steak and potatoes?" one would immediately parse this as choice between

1 steak with no potatoes

2 steak with potatoes

Although strict logic would parse this choice as

1 steak, possibly with potatoes and possibly without potatoes

2 steak, definitely with potatoes,

it is common in English to have the implied negative (i.e. steak with no potatoes) when given a choice where the alternative is a conjunction (i.e. steak with potatoes).

Combinations of Events and the English language If we interpret the doctor's choice with this implied negative, we have:

1 clot with paralysis and no shortness of breath

2 clot with paralysis and shortness of breath

and the first one is much less likely, because it would be odd to have a clot and not have a very common symptom associated with it. The doctor's probability assessment is absolutely correct: both symptoms together are more likely than just one. The "fallacy" arises because the English language is sloppier than mathematical language.

Sum Rule

Now we consider the statements of the form A **or** B. For example, in the card game, what is the fraction of cards that are jacks or are hearts. By counting we get the 13 hearts and 3 more jacks that are not contained in the 13 hearts, or $F\,(\text{jack } \textbf{or } ♥) = \frac{13+3}{52} = 16/52$. Now, if we tried to separate the terms, and do:

$$F\,(\text{jack}) + F\,(♥) = \frac{4}{52} + \frac{13}{52} = \frac{17}{52}$$

then we get a number that is too big! It is too big because we've double-counted the jack of hearts. Adjusting for this, by subtracting one copy of this fraction, we get

$$F\,(\text{jack}) + F\,(♥) - F\,(\text{jack } \textbf{and } ♥) = \frac{4}{52} + \frac{13}{52} - \frac{1}{52} = \frac{16}{52} = F\,(\text{jack } \textbf{or } ♥)$$

In general

Sum Rule

$$P(A \textbf{ or } B) = P(A) + P(B) - P(A \textbf{ and } B) \tag{1.10}$$

Sum Rule

$P(A \textbf{ or } B) = P(A) + P(B) - P(A \textbf{ and } B)$

Sum Rule for Exclusive Events If two events are *mutually exclusive* the sum rule reduces to

$$P(A \textbf{ or } B) = P(A) + P(B) \tag{1.11}$$

because $P(A \textbf{ and } B) = 0$ for such events.

Sum Rule for Exclusive Events If two events are *mutually exclusive* the sum rule reduces to

$P(A \textbf{ or } B) = P(A) + P(B)$

because $P(A \textbf{ and } B) = 0$ for such events.

So the probability of rolling a 1 or a 2 on one die is 2/6.

One more variant on the Sum Rule is where we have 3 propositions. It can be a bit tedious to write it all out, but the end result looks a lot like the original Sum Rule. All we do is break up the terms in pieces, and then apply the Sum Rule to each piece.

$$\begin{aligned}
P(A \textbf{ or } B \textbf{ or } C) &= P(A \textbf{ or } [B \textbf{ or } C]) \\
&= P(A) + P(B \textbf{ or } C) - P(A \textbf{ and } [B \textbf{ or } C]) \\
&= P(A) + P(B) + P(C) - P(B \textbf{ and } C) - \\
&\quad P(A \textbf{ and } B \textbf{ or } A \textbf{ and } C) \\
&= P(A) + P(B) + P(C) - P(B \textbf{ and } C) - \\
&\quad [P(A \textbf{ and } B) + P(A \textbf{ and } C) - \\
&\quad P(A \textbf{ and } B \textbf{ and } A \textbf{ and } C)]
\end{aligned}$$

which leads finally to

Sum Rule for Three Events

$$\begin{aligned}
P(A \textbf{ or } B \textbf{ or } C) &= P(A) + P(B) + P(C) - \\
&\quad P(A \textbf{ and } B) - P(B \textbf{ and } C) - P(A \textbf{ and } C) + \\
&\quad P(A \textbf{ and } B \textbf{ and } C)
\end{aligned} \tag{1.12}$$

Sum Rule for Three Events

$$\begin{aligned}
P(A \textbf{ or } B \textbf{ or } C) &= P(A) + P(B) + P(C) - \\
&\quad P(A \textbf{ and } B) - \\
&\quad P(B \textbf{ and } C) - \\
&\quad P(A \textbf{ and } C) + \\
&\quad P(A \textbf{ and } B \textbf{ and } C)
\end{aligned}$$

In words, when you're looking for the sum of several events, we add the probabilities (i.e. $P(A) + P(B) + P(C)$), then subtract the double counting (i.e. $P(A \textbf{ and } B)$) as before. Finally, we need to add back in the *triple count* (i.e. $P(A \textbf{ and } B \textbf{ and } C)$) because it was taken out too many times with the double count. The accounting here can be somewhat prone to error, but the concepts are always the same: when you add probabilities of events, say A and B, together the term $P(A)$ includes the probability of both $P(A \textbf{ and } B)$ and the term $P(B)$ includes the probability of both $P(A \textbf{ and } B)$, so you've included that probability twice and need to subtract one of them to balance the books. Likewise (although it is harder to show), the first six terms in Equation 1.12 end up subtracting one too many copies of $P(A \textbf{ and } B \textbf{ and } C)$, and we need to add one in at the end.

Marginalization

Another consequence of the sum rule and the product rule is a process called *marginalization*.

EXAMPLE 1.5 *Marginalization and Card Suit*

Imagine we have a number of conditional statements, like:

$$
\begin{aligned}
P(\text{jack}|\heartsuit) &= \frac{1}{13} \\
P(\text{jack}|\diamondsuit) &= \frac{1}{13} \\
P(\text{jack}|\spadesuit) &= \frac{1}{13} \\
P(\text{jack}|\clubsuit) &= \frac{1}{13}
\end{aligned}
$$

but we are interested in just the probability of drawing a jack, regardless of the suit, or in our notation

$$P(\text{jack})$$

The marginalization procedure for this problem looks like:

$$
\begin{aligned}
P(\text{jack}) &= \overbrace{\begin{matrix} P(\text{jack}|\heartsuit) \times P(\heartsuit) + \\ P(\text{jack}|\diamondsuit) \times P(\diamondsuit) + \\ P(\text{jack}|\spadesuit) \times P(\spadesuit) + \\ P(\text{jack}|\clubsuit) \times P(\clubsuit) \end{matrix}}^{\text{all possibilities}} \\
&= \frac{1}{13} \times \frac{1}{4} + \frac{1}{13} \times \frac{1}{4} + \frac{1}{13} \times \frac{1}{4} + \frac{1}{13} \times \frac{1}{4} \\
&= \frac{4}{52}
\end{aligned}
$$

Marginalization If we have a complete set of conditional statements, like

$$P(A|B_1)$$
$$P(A|B_2)$$
$$P(A|B_3)$$
$$P(A|B_4)$$
$$\vdots$$

then the *unconditional* probability is found by *marginalizing* over all possible values of the conditional events, like

$$P(A) = \overbrace{P(A|B_1)P(B_1) + P(A|B_2)P(B_2) + P(A|B_3)P(B_3) + \cdots}^{\text{all possible } Bs} \quad (1.13)$$

Marginalization If we have a complete set of conditional statements, like

$P(A|B_1), P(A|B_2), P(A|B_3), P(A|B_4), \cdots$

then the *unconditional* probability is found by *marginalizing* over all possible values of the conditional events, like

$$P(A) = \overbrace{P(A|B_1)P(B_1) + P(A|B_2)P(B_2) + \cdots}^{\text{all possible } Bs}$$

Bayes' Rule

One of the most consequential rules of probability is what is known as Bayes' Rule, sometimes called Bayes' Theorem. We will use this rule throughout this book, and see its many applications. It comes as a direct result of the product rule (Equation 1.8)

$$P(A \text{ and } B) = P(A|B)P(B) = P(B|A)P(A)$$

In the 1700's Reverend Bayes proved a special case of this rule, and rediscovered in the general form by Pierre-Simon Laplace. Laplace then applied the rule in a large range of problems from geology, astronomy, medicine, and jurisprudence.

Rearranging, we get

Bayes' Rule

$$P(A|B) = \frac{P(B|A)P(A)}{P(B)} \quad (1.14)$$

Bayes' Rule

$$P(A|B) = \frac{P(B|A)P(A)}{P(B)}$$

We can verify this again with the intuitions we have in the simple card game.

EXAMPLE 1.6 *What is the probability of drawing a jack, knowing that you've drawn a face card?*

In terms of fractions, this should be $F(\text{jack}|\text{face card}) = 4/12 = 1/3$. Applying Bayes' Rule to the fractions we get:

$$\begin{aligned} F(\text{jack}|\text{face}) &= \frac{F(\text{face}|\text{jack}) \times F(\text{jack})}{F(\text{face})} \\ &= \frac{\frac{4}{4} \times \frac{4}{52}}{\frac{12}{52}} = \frac{4}{12} = \frac{1}{3} \end{aligned}$$

Although this calculation is true, it isn't particularly enlightening. It is nicer to cast the problem back into probability terms, rather than fractions, and compare the probability of drawing a jack to the probability of the same thing (i.e. drawing a jack) *given* that we know that we've drawn a face card. This is

$$P(\text{jack}) = \frac{1}{13}$$
$$P(\text{jack}|\text{face card}) = \frac{1}{3}$$

This comparison highlights what Bayes' Rule represents: learning. When you are asked what the probability of drawing a jack, from the knowledge of the simple card game, you calculate the value of 1/13. Once you learn that you drew a face card, you update your knowledge to include that information, and modify your probability assignments to reflect this. This leads to an increased chance of the card being a jack.

All of learning is simply updating ones beliefs given the data. The data may be words in a book, the results of an experiment, a conversation with another person, etc... The strength of our beliefs are not often thought of in mathematical terms, but you are doing the math of probabilities whenever you are weighing the strength of your beliefs. Thus, the probabilistic rule - Bayes' rule - for updating beliefs given data is really the quantitative specification of learning. One can use it *qualitatively* as well, which is often useful in fields such as history where the data do not tend to be quantitative.

In a nutshell, Bayes' Rule represents learning:

$$\text{Initial Belief} + \text{New Data} \rightarrow \text{Improved Belief}$$

It is used in science to infer causes from effects, and can thus be written

$$P(\text{cause}|\text{effect}) = \frac{P(\text{effect}|\text{cause}) \times P(\text{cause})}{P(\text{effect})}$$

To infer the probability of a particular cause, given the events you observe in the world, you first have to know the probability of the cause itself (i.e. rarer causes will reduce the *prior* probability), and how likely that the cause you're looking at could have produced the effects you've observed. These two items are the $P(\text{cause})$ and $P(\text{effect}|\text{cause})$ terms, respectively. The entire calculation is scaled by $P(\text{effect})$ which is *all of the other ways that the effects could have been produced by other causes.*. Thus, it is not enough to show that giving a particular medicine is followed by the symptoms disappearing to establish that the medicine was the likely cause of the symptoms disappearing. You have to calculate what other possible causes could have had those effects, such as the normal functioning of the immune system or the placebo effect. This is why carefully controlled studies are necessary, to eliminate all of the other possible causes and to determine the true cause of the effects observed.

We will spend large portions of several chapters on Bayes' Rule, to explore its long-ranging consequences.

1.5 *Venn Mnemonic for the Rules of Probability*

It is often useful to have a picture to represent the mathematics, so that it is easier to remember the equations and to understand their meaning. It is common to use what is called a Venn Diagram to represent probabilities in an intuitive, graphical way. The idea is that probabilities are represented as the *fractional area* of simple geometric shapes. We can then find a picture representation of each of the rules of probability. We start by looking at a sample Venn Diagram, in Figure 1.2.

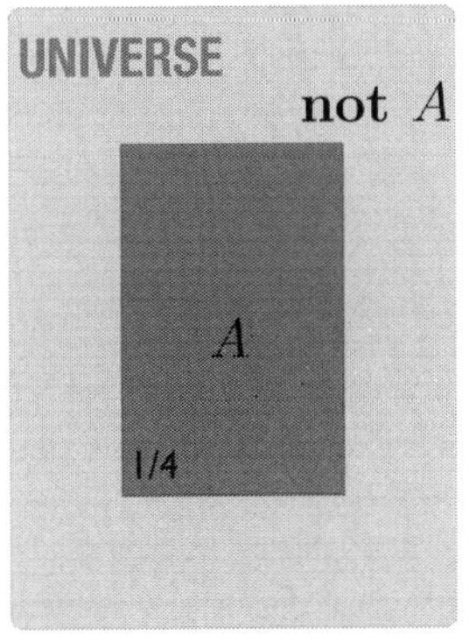

Figure 1.2: Venn diagram of a statement, A, in a *Universe* of all possible statements. It is customary to think of the area of the *Universe* to be equal to 1 so that we can treat the actual areas as fractional areas representing the probability of statements like $P(A)$. In this image, A takes up 1/4 of the *Universe*, so that $P(A) = 1/4$. Also shown is the negation rule. $P(A) + P(\textbf{not } A) = 1$ or "inside" of A + "outside" of A adds up to everything.

The fractional area of the rectangle A represents the probability $P(A)$, and can be thought of as a probability of one of the statements we've explored, such as $P(\heartsuit)$. This diagram is strictly a mnemonic, because the individual points on the diagram are not properly defined. The diagram in Figure 1.2 also represents the Negation Rule (Equation 1.7),

$$P(A) + P(\textbf{not } A) = 1$$

In the diagram it is easy to see that the sum of the areas inside of A (i.e. 1/4) and outside of A (i.e. 3/4) cover the entire area of the *Universe* of statements, and thus add up to 1.

Figure 1.3 shows the diagram which can help us remember the sum and product rules. The Sum Rule (Equation 1.10)

$$P(A \textbf{ or } B) = P(A) + P(B) - P(A \textbf{ and } B)$$

is represented in the total area occupied by the rectangles A and B, and makes up all of A (i.e. 1/4) and the half of B sticking out (i.e. 1/8-1/16=1/16) yielding $P(A \text{ or } B) = 5/16$. This is also the area of each added up (1/4+1/8), but subtracting the intersection (1/16) because otherwise it is counted twice. Adding the areas this way directly parallels the Sum Rule.

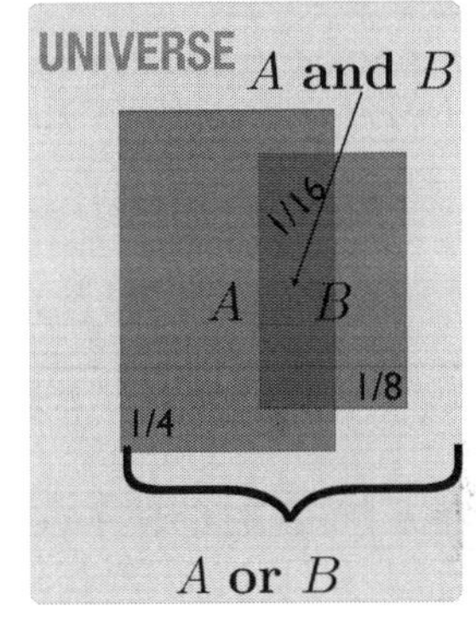

Figure 1.3: Venn diagram of the sum and product. The rectangle B takes up 1/8 of the *Universe*, and the rectangle A takes up 1/4 of the *Universe*. Their overlap here is 1/16 of the *Universe*, and represents $P(A \text{ and } B)$. Their total area of 5/16 of the *Universe* represents $P(A \text{ or } B)$.

Conditional probabilities, like those that come into the Product Rule (Equation 1.8) and Bayes Rule (Equation 1.14) are a little more challenging to visualize. In Figure 1.4, $P(A|B)$ is represented by the fraction of the darker area (which was originally part of A) compared not to the *Universe* but to the area of B, and thus represents $P(A|B) = 1/2$. In a way, it is as if the conditional symbol, "|," defines the *Universe* with which to make the comparisons. On the left of Figure 1.4, the same darker area that was originally part of B represents $P(B|A)$ making up 1/4 of the area of A. Thus $P(B|A) = 1/4$. The Product Rule (Equation 1.8) then follows,

$$P(A \text{ and } B) = \underbrace{P(A|B)}_{1/2}\,\underbrace{P(B)}_{1/8} = \underbrace{P(B|A)}_{1/4}\,\underbrace{P(A)}_{1/4} = \frac{1}{16}$$

We can further see the special case of mutually exclusive statements shown in Figure 1.5. The Sum Rule for Exclusive Events (Equation 1.11) is simply the sum of the two areas because there is no overlap

$$P(A\ \textbf{or}\ B) = P(A) + P(B)$$

Further, it is straightforward to see from this diagram the following properties for mutually exclusive events

$$\begin{aligned} P(A\ \textbf{and}\ B) &= 0 \\ P(A|B) &= 0 \\ P(B|A) &= 0 \end{aligned}$$

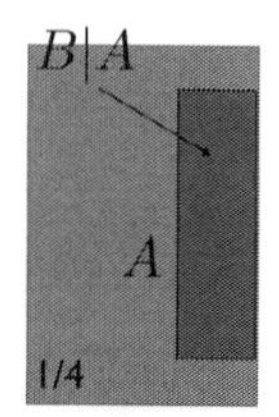

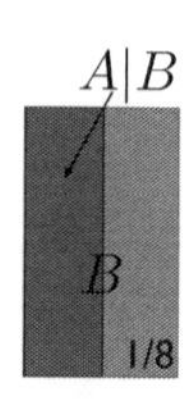

Figure 1.4: Venn diagram of conditional probabilities, $P(A|B)$ and $P(B|A)$. (Right) $P(A|B)$ is represented by the fraction of the darker area (which was originally part of A) compared not to the *Universe* but to the area of B, and thus represents $P(A|B) = 1/2$. In a way, it is as if the conditional symbol, "|," defines the *Universe* with which to make the comparisons. (Left) Likewise, the same darker area that was originally part of B represents $P(B|A)$ which makes up 1/4 of the area of A. Thus $P(B|A) = 1/4$.

1.6 *Lessons from Bayes' Rule - A First Look*

Bayes' Rule is the gold standard for all statistical inference. It is a mathematical theorem, proven from fundamental principles. It structures all inference in a systematic fashion. However, it can be used without doing any calculations, as a guide to qualitative inference. Some of the lessons which are consequences of Bayes' Rule are listed here, and will be noted throughout this text in various examples.

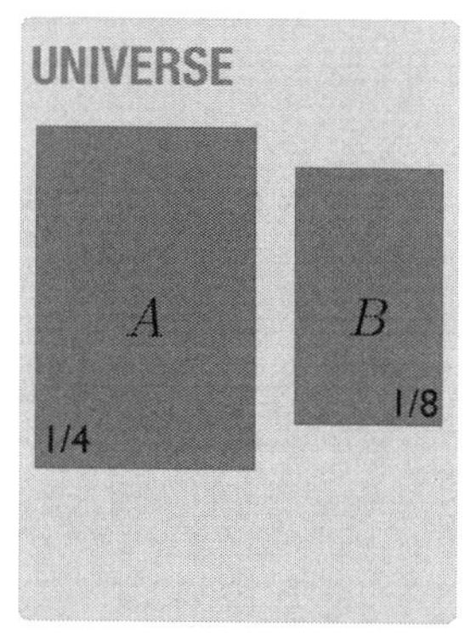

Figure 1.5: Venn diagram of mutually exclusive statements. One can see that $P(A\ \textbf{and}\ B) = 0$ (the overlap is zero) and $P(A\ \textbf{or}\ B) = P(A) + P(B)$ (the total area is just the sum of the two areas)

- Confidence in a claim should scale with the evidence for that claim
- Ockham's razor, which is the philosophical idea that simpler theories are preferred, is a consequence of Bayes' Rule when comparing models of differing complexity.
- Simpler means fewer adjustable parameters
- Simpler also means that the predictions are both *specific* and not *overly plastic*. For example, a hypothesis which is consistent with the observed data, and also be consistent if the data were the opposite would be overly plastic.
- Your inference is only as good as the hypotheses (i.e. models) that you consider.
- Extraordinary claims require extraordinary evidence.[7]
- It is better to explicitly display your assumptions rather than implicitly hold them.
- It is a *good thing* to update your beliefs when you receive new information.
- Not all uncertainties are the same.

[7] Carl Sagan. *Demon-Haunted World: Science as a Candle in the Dark*. Random House LLC, 1996

There is not a universal agreement for the translation of numerical probability values to qualitative terms in English (i.e. highly unlikely, somewhat unlikely, etc...). One rough guide is shown in Table 1.1. I will be following this convention throughout the book, but realize that the specific probability distinctions are a bit arbitrary.

term	probability
virtually impossible	1/1,000,000
extremely unlikely	0.01 (i.e. 1/100)
very unlikely	0.05 (i.e. 1/20)
unlikely	0.2 (i.e. 1/5)
slightly unlikely	0.4 (i.e. 2/5)
even odds	0.5 (i.e. 50-50)
slightly likely	0.6 (i.e. 3/5)
likely	0.8 (i.e. 4/5)
very likely	0.95 (i.e. 19/20)
extremely likely	0.99 (i.e. 99/100)
virtually certain	999,999/1,000,000

Table 1.1: Rough guide for the conversion of qualitative labels to probability values.

2 *Applications of Probability*

In this chapter we go through a number of examples of the uses of probability, and present several useful mathematical tools along the way.

2.1 *The Birthday Problem*

This is a famous problem in probability[1], which we address here in stages. We introduce a simple version, and make it more complex in steps until we can tackle the general problem.

[1] F. Mosteller. *Fifty challenging problems in probability with solutions.* Dover Pubns, 1965

Two People on April 3

EXAMPLE 2.1 *Let's imagine we have the case where two people meet on the street. What is the probability that they both have April 3 as their birthday?*

This can be solved with a straightforward application of the product rule, Equation 1.8 on page 36.

$$\begin{aligned} A &\equiv \text{Person 1 has a birthday on, say, April 3} \\ B &\equiv \text{Person 2 has a birthday on, say, April 3} \\ P(A \text{ and } B) &= P(A|B)P(B) \end{aligned}$$

Each of these terms can be calculated. Firstly, $P(A|B)$ is the probability that person 1 has a certain birthday given that person 2 has the same birthday. However, knowing the birthday of the second person doesn't tell us anything about the birthday of the first person, thus they are *independent* and $P(A|B) = P(A)$.

Secondly, the probability of having any particular birthday is simply $P(A) = 1/365$. Finally, we have

This is the simplest assumption - that each day is equally likely to be born on. However, this is probably not true - there are some days that are more likely than others. In addition, once you start including February 29, then things obviously change.

$$\begin{aligned} A &\equiv \text{Person 1 has a birthday on, say, April 3} \\ B &\equiv \text{Person 2 has a birthday on, say, April 3} \\ P(A \text{ and } B) &= \frac{1}{365} \times \frac{1}{365} = \frac{1}{133,225} = 0.0000075 \end{aligned}$$

which is extremely unlikely (see Table 1.1 on page 45)!

Two People

EXAMPLE 2.2 *Two people meet on the street, and we ask what is the probability that they both have the same birthday?*

How is this different than the previous question, where we specified which birthday they had? Our intuition immediately suggests that this probability must be *higher* than the previous one, because there are more possibilities - rather than April 3, they could be born on January 1 or May 3 or any other day. Using our notation we have the following definitions:

In all of these examples we are not considering leap days, which occur approximately once every four years. These extra days do not change any of the qualitative results, and really only serve as a small extra correction to any analysis. However, it does add a fair amount of bookkeeping with very little increase in enlightenment, so we choose to avoid this problem in our examples.

$$\begin{aligned} C_1 &\equiv \text{Person 1 and Person 2 both have a birthday on January 1} \\ C_2 &\equiv \text{Person 1 and Person 2 both have a birthday on January 2} \\ &\vdots \\ C_{365} &\equiv \text{Person 1 and Person 2 both have a birthday on December 31} \end{aligned}$$

and the probability we are looking for is

$$P(C_1 \text{ or } C_2 \text{ or } \cdots \text{ or } C_{365})$$

In this situation we can note that these are *exclusive* statements. For example, it can't be true that both C_1 *and* C_2 are true - you can't have more than one birthday. Thus, the Sum Rule (Equation 1.10 on page 1.10) reduces to the Limited Sum Rule (Equation 1.11). Further, each term in that rule is the same

$$P(C_1) = P(C_2) = \cdots = P(C_{365}) = \frac{1}{365} \times \frac{1}{365}$$

so we have

$$\begin{aligned} P(C_1 \text{ or } C_2 \text{ or } \cdots \text{ or } C_{365}) &= \\ &\underbrace{\left(\frac{1}{365} \times \frac{1}{365}\right) + \left(\frac{1}{365} \times \frac{1}{365}\right) + \cdots + \left(\frac{1}{365} \times \frac{1}{365}\right)}_{\text{365 terms, one for each day}} \\ &= \frac{1}{365} = 0.0027 \end{aligned}$$

Another way to think of this is to imagine that person 1 randomly "chooses" their birthday, D_1, and person 2 randomly "chooses" their birthday, D_2, and then they compare to see if the days are the same, or $D_1 = D_2$. In general, we can think of the problem broken up in this way:

Here we find another example of the general requirement that equivalent states of knowledge give rise to equivalent probability assignments. In this case it means that if there is more than one way to arrive at a conclusion, they each must give the same answer. We can then choose the way that is easiest to calculate, simply out of convenience.

$$\begin{aligned} P(D_1 = D_2) &= \\ &P\begin{pmatrix} D_1 \text{ is a specific day } \textbf{and} \\ D_2 \text{ is the same day} \end{pmatrix} \times \begin{pmatrix} \text{number of possible} \\ \text{specific days} \end{pmatrix} \end{aligned}$$

In this way, we get

$$
\begin{aligned}
P(D_1 = D_2) &= \left(\frac{1}{365} \times \frac{1}{365}\right) \times (365) \\
&= \frac{1}{365} = 0.0027
\end{aligned}
$$

which is extremely unlikely (see Table 1.1 on page 45), but not nearly as unlikely as them both having the same April 3 birthday.

Three People

EXAMPLE 2.3 *What is the probability that three random people have the same birthday?*

Going through the same logic, we have

$$
\begin{aligned}
P(D_1 = D_2 = D_3) &= \left(\frac{1}{365} \times \frac{1}{365} \times \frac{1}{365}\right) \times 365 \\
&= \frac{1}{133,225} = 0.0000075
\end{aligned}
$$

which is even more extremely unlikely (see Table 1.1 on page 45) then the previous two-person example. It is interesting to note that this is the same answer we received when we asked for the probability of two people with a *specific* birthday. One can think of the the three people having the same, unspecified, birthday in the following way if it helps. The first person's birthday specifies the necessary birthday for the other two, so it is the same as the case where we specify a single birthday for two people.

Two People...Out of Three

Usually, we don't have a situation where we have random people meeting and all agreeing on birthdays. What we have is a group of people talking, and two people in the group end up saying "Hey, my birthday is April 3 too!" This is quite a bit different, and leads to some unintuitive consequences. Let's go through the situation with three people, and we ask the question

EXAMPLE 2.4 *What is the probability that* at least two *have the same birthday?*

Writing this out we get (somewhat messily)

$$
\begin{aligned}
&P(\text{at least two out of three have the same birthday}) = \\
&= P(\text{exactly 2 the same } \textbf{or} \text{ exactly 3 the same}) \\
&= P(\text{exactly 2 the same}) + \\
&\quad \underbrace{P(\text{exactly 3 the same})}_{\left(\frac{1}{365}\right)^3 \times 365} - \underbrace{P(\text{exactly 2 } \textbf{and} \text{ exactly 3 the same})}_{0}
\end{aligned}
$$

Writing the possibilities out like this is quite tedious, and can lead to errors. Directly after this calculation we find an equivalent, and much easier, way of writing the same calculation. However, it is important to note that all ways of writing the same information must lead to the same answer.

The term $P(\text{exactly 2 the same})$ can be broken up like

$$\begin{aligned} P(\text{exactly 2 the same}) &= P(\text{a specific 2 are the same}) \times \begin{pmatrix} \text{number of} \\ \text{possibilities of} \\ \text{2 the same} \end{pmatrix} \\ &= P(D_1 = D_2 \text{ and } \textbf{not } D_1 = D_3) \times \begin{pmatrix} \text{number of} \\ \text{possibilities of} \\ \text{2 the same} \end{pmatrix} \end{aligned}$$

Applying the product rule we get

I'm sure you're wishing for the easier way about now...it's coming in Example 2.5.

$$\begin{aligned} & P(\text{exactly 2 the same}) = \\ &= P(D_1 = D_2 \text{ and } \textbf{not } D_1 = D_3) \times \begin{pmatrix} \text{number of} \\ \text{possibilities of} \\ \text{2 the same} \end{pmatrix} \\ &= P(D_1 = D_2 | \textbf{not } D_1 = D_3) P(\textbf{not } D_1 = D_3) \times \begin{pmatrix} \text{number of} \\ \text{possibilities of} \\ \text{2 the same} \end{pmatrix} \\ &= \underbrace{P(D_1 = D_2)}_{\frac{1}{365}} \underbrace{P(\textbf{not } D_1 = D_3)}_{\frac{364}{365}} \times \begin{pmatrix} \text{number of} \\ \text{possibilities of} \\ \text{2 the same} \end{pmatrix} \end{aligned}$$

Noting that there are 3 ways of getting a specific 2 the same, we obtain for this single term

These 3 ways are "person 1 and 2 match", "person 1 and 3 match", "person 2 and 3 match."

$$P(\text{exactly 2 the same}) = \frac{1}{365} \times \frac{364}{365} \times 3$$

Putting it all together we have

$$\begin{aligned} & P(\text{at least two out of three have the same birthday}) = \\ &= P(\text{exactly 2 the same } \textbf{or} \text{ exactly 3 the same}) \\ &= \frac{1}{365} \times \frac{364}{365} \times 3 + \left(\frac{1}{365}\right)^3 \times 365 \\ &= 0.0082 \end{aligned}$$

EXAMPLE 2.5 *What is the probability that* at least two *have the same birthday? A clever shortcut.*

A clever way of rethinking this problem, which significantly reduces the calculations, is found by asking the following question: in a group of people, what is the probability that *none* of the people have the same birthday? This can be approached in a step-wise fashion. Person 1 "chooses" a birthday, out of 365 they have all 365 possibilities. Person 2 "chooses" their birthday, with probability $P = 364/365$ of *not* being the same as Person 1. Person 3 now has 363 "choices" out of 365 to avoid both other birthdays, etc... So the probability of using this process and getting to Person 3 and not have any overlapping birthdays is simply

$$P(\text{none the same in 3 people}) = \frac{365}{365} \times \frac{364}{365} \times \frac{363}{365}$$

Now, if we're interested in the probability that at least two are the same, then this is the exact *opposite* of the probability that none are the same. Using the Negation Rule (Equation 1.7 on page 35) we have

$$P\begin{pmatrix}\text{none the same}\\ \text{in 3 people}\end{pmatrix} + P\begin{pmatrix}\textbf{not} \text{ ``none}\\ \text{the same in 3}\\ \text{people''}\end{pmatrix} = 1$$

$$P\begin{pmatrix}\text{none the same}\\ \text{in 3 people}\end{pmatrix} + P\begin{pmatrix}\text{at least 2 the}\\ \text{same in 3}\\ \text{people}\end{pmatrix} = 1$$

which leads to

$$\begin{aligned} P\begin{pmatrix}\text{at least 2 the}\\ \text{same in 3}\\ \text{people}\end{pmatrix} &= 1 - P\begin{pmatrix}\text{none the same}\\ \text{in 3 people}\end{pmatrix} \\ &= 1 - \frac{364}{365} \times \frac{363}{365} \\ &= 0.082 \end{aligned}$$

Two People...Out of Thirty

EXAMPLE 2.6 *When you have a group of 30 people, like students in a classroom, and you ask what the probability of finding two in the room with the same birthday, would your intuition say it is greater or less than 50%?*

Many people find that their intuition suggests reasonably strongly that it would be less than 50%. We can now do this problem quite easily, and we find that our intuition does not match. Following the same procedure as with 3 people, we imagine each person "choosing" their birthday with a dwindling selection as we go on to avoid "choosing" one that has already been taken. The probability that no one in the room has the same birthday as any other is

We've often used our intuition to verify the result, but now we've reached a state where the problems get subtle enough that our intuition fails. It is good to use ones' intuition on the "easy" problems, but now that we've established the process we can tackle problems where our intuition is not good enough to confirm a result.

$$\begin{aligned} P(\text{none the same in 30 people}) &= \underbrace{\frac{365}{365} \times \frac{364}{365} \times \frac{363}{365} \times \cdot \times \frac{335}{365}}_{\text{30 terms}} \\ &= 0.27 \end{aligned}$$

So the probability of having at least 2 people in the room having the same birthday is

$$P\begin{pmatrix}\text{at least 2 the}\\ \text{same in 30}\\ \text{people}\end{pmatrix} = 1 - 0.27$$

$$= 0.73$$

which is 73%! Compare this likely outcome to the extremely rare outcome of having two random people having matched birthdays, from page 48. See Figure 2.1 to see a plot of this unintuitive observation.

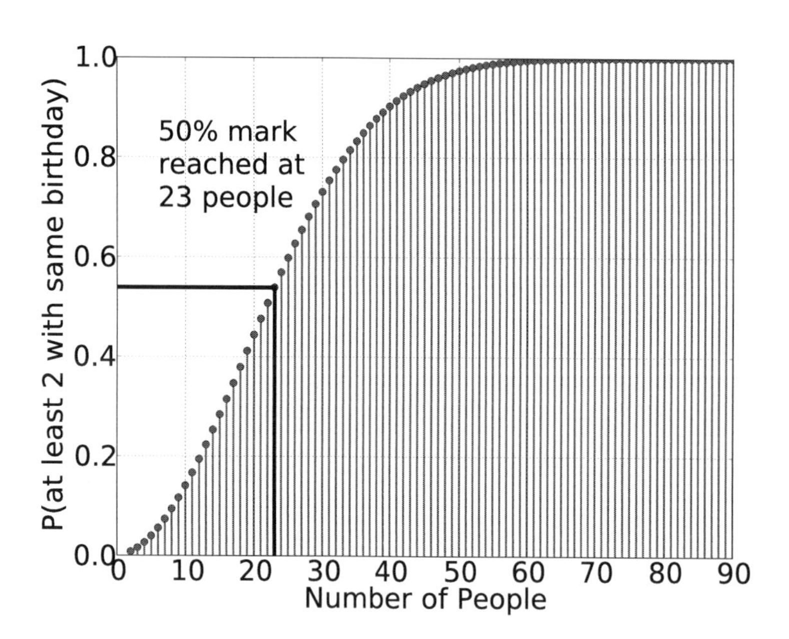

Figure 2.1: Probability of having at least two people in a group with the same birthday depending on the number of people in the group. The 50% mark is exceeded once the group size exceeds 23 people.

2.2 *The Lottery Problem or Rare Things Are Common*

This problem is identical to the birthday problem mathematically,with the only difference that the probability numbers are much smaller and the number of participants is much larger. We start with a story about someone winning the lottery *twice* in the same day![2]

[2] Calif. couple wins two lotteries in one day, 2002. URL `http://abcnews.go.com/US/story?id=90981`

> Can you imagine winning the lottery twice in one day?
>
> Angelo and Maria Gallina don't have to imagine Ñ they hit twice on Nov. 20.
>
> The married couple from Belmont, Calif., had separately bought tickets in two different California state lottery games, and both could hardly believe their eyes as all 11 winning numbers over two games came up....Before taxes, their winnings amounted to $126,000 for the Fantasy 5 and 17$ million for the SuperLotto Plus, according to The Associated Press....Orkin arrived at the number by multiplying the roughly 41-million-to-one odds of winning the SuperLotto game and the 575,000-to-one odds of winning the Fantasy 5 game to arrive at odds of 23,575,000,000,000-to-one.

Pretty amazing! That's something like

$$P(\text{winning two tickets}) = \frac{1}{2\times 10^{13}} \sim 5\times 10^{-14} \qquad (2.1)$$

which truly is quite improbable as a single event, but is it truly an improbable event to happen *somewhere*? The assumption stated in the quote is that only two tickets were purchased. We all know that many lottery tickets are purchased daily, which should increase the chance that *somewhere* this will occur. Even this winning couple purchased tickets every day for 20 years before winning this.

Like the birthday problem, you have to set up the problem in the negative, and as what the probability of no one winning two lotteries. If we assume 5 million people playing daily for 20 years, this probability is:

$$P\left(\begin{array}{l|l}\text{no one} & \text{5 million} \\ \text{winning} & \text{plays daily} \\ \text{two tickets} & \text{for 20} \\ & \text{years}\end{array}\right) = \left(1-\frac{1}{2\times 10^{13}}\right)^{5\times 10^6\times 365\times 20} \qquad (2.2)$$

$$\sim (1-5\times 10^{-14})^{5\times 10^6\times 365\times 20}$$

$$= 0.998 \qquad (2.3)$$

yielding a 0.2% chance of this happening sometime in those 20 years - still pretty rare, but not outrageously so. If we further imagine that this is occurring across the 50 states, this increases to 10% chance of this happening sometime in those 20 years. If we further imagine that there are as many as 5 different lotteries (there are usually more) that could be played per state, this jumps up to 40%.

What we see as an initially *highly* unlikely event starts to become *likely* and in fact *common* when considering all of the possible ways that event could be produced.

2.3 *Monte Hall Problem*

One of the most popular probability problems is called the Monte Hall problem, and is based on the television game show "Let's Make a Deal."[3] It can take on many forms, but a common form is as follows[4]

[3] S. Selvin. A problem in probability. *American Statistician*, 29(1):67, 1975

[4] M. Vos Savant. Ask Marilyn [column]. *Parade Magazine*, page 16, 1990

EXAMPLE 2.7 *Suppose you're on a game show, and you're given the choice of three doors: behind one door is a car; behind the others, goats. You pick a door, say No. 1 (but the door is not opened), and the host, who knows what's behind the doors, opens another door, say No. 3, which has a goat. He then says to you, "Do you want to changed your choice to door No. 2?" Is it to your advantage or disadvantage to switch your choice, or does it matter whether you switch your choice or not?*

The result is that *it is always better to switch*, where the probability of getting the car moves up from 1/3 to 2/3 by switching! Because this problem is particularly unintuitive, we will break it up into smaller pieces. The critical aspect of this is that a *change in our assignment of probability to an event must be somehow tied to a change in our information about that event*. In order to understand the problem, we must then understand where the extra information is coming from.

Most people will state that, because we are left with 2 choices, it must be 50-50. However there is added information in the system which moves us from knowing *nothing* about the two choices (i.e. 50-50 chance) to knowing *a little bit more* about the two choices (i.e. not 50-50 chance).

We will step up to the full problem listed, but for now we explore some simpler versions of the problem.

Two Doors with Information

EXAMPLE 2.8 *Imagine we have a game with two doors: Behind one door is a car; behind the other is a goat. You pick a door, say No. 1 (but the door is not opened), and the host, who knows what's behind the doors, says that there is a 90% chance that the car is behind door No. 2. Is it to your advantage to switch your choice?*

Initially there is a two-door choice, with no information about either choice, so we assign equal probabilities to the choices: $P(\text{car behind No. 1}) = P(\text{car behind No. 2}) = 0.5$ (i.e. a 50-50 chance). After the host gives information, this changes. Although this is still a two-door choice, it is no longer a 50-50 chance. By having a knowledgable person give you information suddenly changes the situation to a 10-90 chance, and it is much better for you to switch.

What if the host were a little less direct? Perhaps something like

EXAMPLE 2.9 *The host, who knows what's behind the doors, points to a door, choosing the correct door 90% of the time and the incorrect one 10%. You pick a door, say No. 1, and the host points to door No. 2. Is it to your advantage to switch your choice?*

This amounts to an identical situation as the previous one - the host is giving you correct information 90% of the time, and we are in a much better position switching.

Three Doors with Information

We return to the three-door case with a slight variation

EXAMPLE 2.10 *Suppose you're on a game show, and you're given the choice of three doors: Behind one door is a car; behind the others, goats. You pick a door, say No. 1 (but the door is not opened), and the host, who knows what's behind the doors, says that another door, say No. 3, has a 0% chance of having a car, and that the remaining door (that you haven't chosen - i.e door No. 2) has a 66% of having the car. He then says to you, "Do you want to pick door No. 2?" Is it to your advantage to switch your choice?*

In this case, switching to door No. 3 would be ridiculous - we know the car isn't there, because the (honest) host knows that it is not there. The host also has told us that there is a 66% chance of the car behind door No. 2, and thus we have $P(\text{car behind No. 1}) = 0.34$ and $P(\text{car behind No. 2}) = 0.66$ and it is better to switch to door No. 2.

It isn't the number of choices that is important, it is the information we have about those choices. When you have no information, we assign equal probabilities. When we have information, we can assign non-equal probabilities.

Three Doors Down To Two

Back to our original problem, we have

EXAMPLE 2.11 *Suppose you're on a game show, and you're given the choice of three doors: behind one door is a car; behind the others, goats. You pick a door, say No. 1 (but the door is not opened), and the host, who knows what's behind the doors, opens another door, say No. 3, which has a goat. He then says to you, "Do you want to changed your choice to door No. 2?" Is it to your advantage or disadvantage to switch your choice, or does it matter whether you switch your choice or not?*

The key part is that, no matter what happens,

1 the host *never* opens your door

2 the host *always* opens a door with a goat

Given that your first choice, with three equal probability choices (i.e. you have no information about any of the choices), we expect to be correct only about 33% of the time. If we happened to get lucky with our first choice, then the host has a pick of two doors with goats and has some freedom. If we happened to get unlucky with our first choice (and there is a goat behind it), then the host has *no freedom at all*, because there is only one remaining door with a goat. So, about 66% of the time the host is forced to reveal some of his information, because the door he leaves closed (other than your door) *must* have the car. Thus, 66% of the time the host is *telling you where the car is*, just a little indirectly.

Another way to look at this is to imagine a game with 1000 doors, car behind only one, and the host has to open up 998 doors (not yours and not the prize - if the prize is different than yours). Once you pick, say door number 1, and the host opens every door except door 576, and gives you the opportunity to switch is it a good choice? Of course! Ones intuition realizes that my initial 1/1000 chance of getting it right (and thus have the other door have a goat) is swamped by the 999/1000 chance of getting it wrong, and the host being forced to open every door without the prize.

Formally, we need to involve model comparison, so we postpone this particular analysis until Section 5.2.

2.4 *Weather*

EXAMPLE 2.12 *If the probability that it will rain next Saturday is 0.25 and the probability that it will rain next Sunday is 0.25, what is the probability that it will rain during the weekend?*

First Solution - Independence

If we assume that Sunday and Saturday weather are *independent* then the sum-rule (Section 1.4) applies:

$$
\begin{aligned}
&P(\text{rain Saturday } \textbf{or} \text{ rain Sunday}) = \\
&\quad P(\text{rain Saturday}) + P(\text{rain Sunday}) - P(\text{rain Saturday } \textbf{and} \text{ rain Sunday}) \\
&= P(\text{rain Saturday}) + P(\text{rain Sunday}) - P(\text{rain Saturday}) \times P(\text{rain Sunday}) \\
&= 0.25 + 0.25 - 0.25 \times 0.25 = 0.4375 \qquad (2.4)
\end{aligned}
$$

The diagrams in Figure 1.3 are useful in making this calculation more intuitive, especially the term where we subtract $P(\text{rain Saturday}) \times P(\text{rain Sunday})$ because otherwise we over count the double-rain weekends.

Another way to think of this term can be seen in answering a different question - *what is the total number of weekends with rain?*. Imagine we have, in a year, 200 Saturdays with rain (by simply going through all of the Saturdays and counting them if it rains on that day) and we also have 200 Sundays with rain. If we want to know the number of weekends with rain we can add the Saturdays with rain and the Sundays with rain (coming to four hundred!) and it becomes clear that we've over counted those weekends where it rained both days, and we'd need to subtract those to get a reasonable answer. The same logic applies to the calculation of probabilities.

Second Solution - Correlation

Is it really reasonable that rain on Saturday and Sunday are independent events? Probably not! It's probably the case that knowing that it rained on Saturday, then rain on Sunday is more likely. It may also be that if it *didn't rain* on Saturday then it will be *less likely* for rain on Sunday. So we'd have information possibly like:

$$
\begin{aligned}
P(\text{rain Sunday}|\text{rain Saturday}) &= 0.35 \\
P(\text{rain Sunday}|\text{not rain Saturday}) &= 0.15
\end{aligned}
$$

Knowing this changes the equation as

$$
\begin{aligned}
&P(\text{rain Saturday } \textbf{or} \text{ rain Sunday}) = \\
&= P(\text{rain Saturday}) + P(\text{rain Sunday}) - P(\text{rain Saturday } \textbf{and} \text{ rain Sunday})
\end{aligned}
$$

Notice, however, that we don't have a direct expression for $P(\text{rain Sunday})$ anymore. We only have the *conditional* or *dependent* forms, like $P(\text{rain Sunday}|\text{rain Saturday})$. We can use the marginalization procedure (Equation 1.13 on page 41), and sum over all of the conditional expressions

$$
\begin{aligned}
P(\text{rain Sunday}) &= P(\text{rain Sunday}|\text{rain Saturday})\, P(\text{rain Saturday}) + \\
&\quad P(\text{rain Sunday}|\textbf{not} \text{ rain Saturday})\, P(\textbf{not} \text{ rain Saturday}) \\
&= 0.35 \times 0.25 + 0.15 \times (1 - 0.25) = 0.2
\end{aligned}
$$

and then we have

$$
\begin{aligned}
&= P(\text{rain Saturday}) + P(\text{rain Sunday}) - P(\text{rain Saturday}) \times P(\text{rain Sunday}|\text{rain Saturday}) \\
&= 0.25 + 0.2 - 0.25 \times 0.35 = 0.3625 \qquad (2.5)
\end{aligned}
$$

which makes it *less likely* to rain on the weekend if the Sunday rain is correlated with the Saturday rain (Equation 2.5) than if they are independent (Equation 2.4). Why is that?

One way to think of it is that, although the probability of rain on Sunday is increased due to rain on Saturday, it is more likely that Saturday is not rainy. In those cases, which are more frequent, Sunday is less likely to be rainy as well. When the two days are independent, Sunday's rain is the same probability regardless of Saturday's weather. When they are dependent, then the more often clear Saturday weather makes it a little less likely for the Sunday rain, and thus lowers the chance of weekend rain by a little bit.

2.5 *Claims and Priors*

Doctors' Claims - English Language and Probability

In Section 1.4 we introduced work by Tversky and Kahneman documenting supposed failures in proper reasoning. In the example survey of medical internists, the internists were asked

> Which is more likely: the victim of an embolism (clot in the lung) will experience partial paralysis or that the victim will experience both partial paralysis and shortness of breath?

and 91 percent of the doctors chose that the clot was less likely to cause the rare paralysis rather than to cause the combination of the rare paralysis and the common shortness of breath.

This may not be a failure of reasoning, but a (correct!) failure of the doctors to translate the English language literally into logical language. It is likely that when doctors are asked: "Which is more likely: that the victim of an embolism will experience partial paralysis or that the victim will experience both partial paralysis and shortness of breath?" they interpret it as:

1. someone is *claiming* that the patient has an embolism
2. the patient is *claiming*, or someone has measured, that she has partial paralysis
3. the patient is *claiming*, or someone has measured, that she has shortness of breath

The doctors are separating the analysis of the *claim* of the clot, which is given information, from the other claims. Another way of looking at it is to include the knowledge of the method of reporting. Someone who is reporting information about an ailment will tend to report all of the information accessible to them. By reporting only the paralysis, there are two possibilities concerning the person measuring the symptoms of the patient:

1 they had the means to measure shortness breath in the patient, but there was none

2 they did not have the means to measure shortness of breath

In the first case, the doctor's probability assessment is absolutely correct: both symptoms together are more likely than just one. In the second case, the doctors are also correct: one of the sets of diagnostic results (i.e. just paralysis) is less dependable than the other set (i.e. both symptoms), thus the second one is more likely to indicate a clot or is consistent with the known clot.

It isn't that the doctors are reasoning incorrectly. They are including more information, and doing a more sophisticated inference than the strict, formal, minimalistic interpretation of the statements would lead one to do. This analysis works well for other examples stated in the book *A Drunkard's Walk* by Mlodinow[Mlodinow, 2008], like "Is it more probable that the president will increase federal aid to education or that he or she will increase federal aid to education with funding freed by cutting other aid to states?"

All of this underscores the need to be careful translating statements of probability into plain English and vice versa.

Diverging Opinions

Is it possible to have people informed by the same information, and reasoning properly, to have diverging opinions? It might seem intuitive that people given the same information, reasoning properly, would tend to come to agreement, however this is not always the case. What is interesting is that it turns on the prior probabilities for claims. This example comes from Jaynes, 2003[5]. We have the following piece of information:

[5] E. T. Jaynes. *Probability Theory: The Logic of Science*. Cambridge University Press, Cambridge, 2003. Edited by G. Larry Bretthorst

$$D := \begin{cases} \text{"Mr } N\text{. has gone on TV with a sensational claim} \\ \text{that a commonly used drug is unsafe"} \end{cases}$$

and we have observers A, B, and C with different prior assignments to the reliability of Mr N and of the safety of the drug. These prior assignments may have been the result of previous inference by these observers, in a different context, or possible due to expert knowledge. Observers A and C believe, before the announcement, that the drug is reasonably safe. Observer B does not. We have the probability assignments then:

$$P_A(\text{Safe}) = 0.9$$

$$
\begin{aligned}
P_B(\text{Safe}) &= 0.1 \\
P_C(\text{Safe}) &= 0.9
\end{aligned}
$$

They all agree that if the drug is not safe, then Mr N would announce it, so we have

$$
\begin{aligned}
P_A(D|\text{not Safe}) &= 1 \\
P_B(D|\text{not Safe}) &= 1 \\
P_C(D|\text{not Safe}) &= 1
\end{aligned}
$$

Finally, we have the perceptions from the observers about the reliability of Mr N if the drug is actually safe. In this case, observer A is trusting of Mr N, observer C is strongly distrustful, and observer B is mildly distrustful. By "distrustful" we are referring to the probabilities that Mr N would make the announcement that the drug is unsafe *even if* the drug were actually safe. So we have

$$
\begin{aligned}
P_A(D|\text{Safe}) &= 0.01 \\
P_B(D|\text{Safe}) &= 0.3 \\
P_C(D|\text{Safe}) &= 0.99
\end{aligned}
$$

We want to know how each observer then determines whether the drug is safe, given the announcement, or $P(\text{Safe}|D)$ for each observer.

Applying Bayes' Rule we have

$$
\begin{aligned}
P_A(\text{Safe}|D) &= \frac{P_A(D|\text{Safe})P_A(\text{Safe})}{P_A(D|\text{Safe})P_A(\text{Safe}) + P_A(D|\text{not Safe})P_A(\text{not Safe})} \\
&= \frac{0.01 \cdot 0.9}{0.01 \cdot 0.9 + 1 \cdot 0.1} = 0.083
\end{aligned}
$$

Following the same calculation for the others, we get the observers updating their probability assignments after the announcement, D, as

$$
\begin{aligned}
P_A(\text{Safe}) = 0.9 &\rightarrow P_A(\text{Safe}|D) = 0.083 \\
P_B(\text{Safe}) = 0.1 &\rightarrow P_B(\text{Safe}|D) = 0.032 \\
P_C(\text{Safe}) = 0.9 &\rightarrow P_C(\text{Safe}|D) = 0.899
\end{aligned}
$$

Observer A changed their mind, Observer B had their assessment confirmed a bit, and Observer C barely budged.

Although you'd think that hearing the announcement of the unsafe nature of the drug would have moved all of the probabilities by the same amount, but the information isn't that the drug is unsafe, but the someone is *claiming* that the drug is unsafe. Thus, ones prior information about both the drug and who is making the claim comes into play.

2.6 Adding Dice

EXAMPLE 2.13 *What is the probability of the* sum *of two dice getting a particular value, say, 7?*

In this case, we simply outline every single possibility, and count the fractions. In a more complex case we may need to find a better method of counting, but the idea will be the same.

We find immediately that the probability of getting a sum of 7 is the largest, because there are more arrangements of the two dice which yield a sum of 7 than for any other sum.

Each probability of a particular sum is just the number of arrangements to get that particular sum divided by the total number of arrangements of a two dice (i.e. 36).

All possible results from rolling two dice:

sum	(die 1,die 2)
2	(1,1)
3	(1,2),(2,1)
4	(3,1),(1,3),(2,2)
5	(1,4),(4,1),(3,2),(2,3)
6	(1,5),(5,1),(4,2),(2,4),(3,3)
7	(1,6),(6,1),(5,2),(2,5),(4,3),(3,4)
8	(3,5),(5,3),(6,2),(2,6),(4,4)
9	(5,4),(4,5),(3,6),(6,3)
10	(4,6),(6,4),(5,5)
11	(6,5),(5,6)
12	(6,6)
	(36 arrangments total)

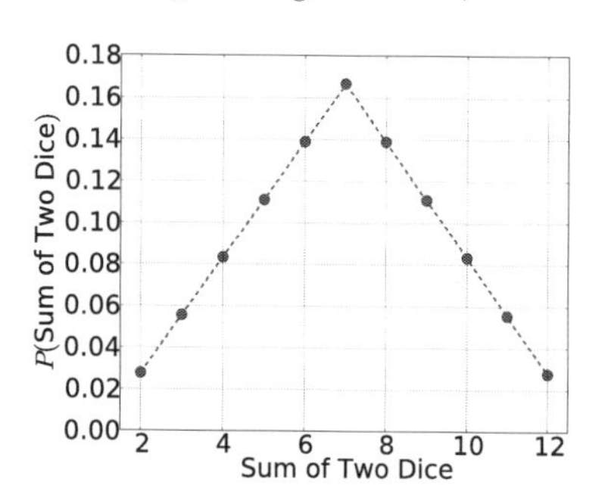

$$
\begin{aligned}
P(2) &= \frac{1}{36} = 0.028 & P(8) &= \frac{5}{36} = 0.139 \\
P(3) &= \frac{2}{36} = 0.055 & P(9) &= \frac{4}{36} = 0.111 \\
P(4) &= \frac{3}{36} = 0.083 & P(10) &= \frac{3}{36} = 0.083 \\
P(5) &= \frac{4}{36} = 0.111 & P(11) &= \frac{2}{36} = 0.055 \\
P(6) &= \frac{5}{36} = 0.139 & P(12) &= \frac{1}{36} = 0.028
\end{aligned}
$$

EXAMPLE 2.14 *What is the probability of rolling a sum* more than 7 *with two dice?*

In our notation this is

$$P(8 \text{ or } 9 \text{ or } 10 \text{ or } 11 \text{ or } 12)$$

which are all *exclusive events*, so we use the *Sum Rule* for exclusive events (Equation 1.11) and obtain

$$
\begin{aligned}
P(8 \text{ or } 9 \text{ or } 10 \text{ or } 11 \text{ or } 12) &= P(8) + P(9) + P(10) + P(11) + P(12) \\
&= 0.139 + 0.111 + 0.083 + 0.055 + 0.028 \\
&= 0.416
\end{aligned}
$$

EXAMPLE 2.15 *What is the probability of rolling various sums with two dice* each with 20 sides*?*

20-sided dice are common in some kinds of games, and provide a nice alternative to the standard 6-sided variety. The figure comparing the 6-sided and 20-sided dice can be see in in Figure 2.2 on page 61.

2.7 Cancer and Probability

Here we explore an example that we cover again, in a slightly different way, in Section 5.1 on page 103. Imagine we have a population of

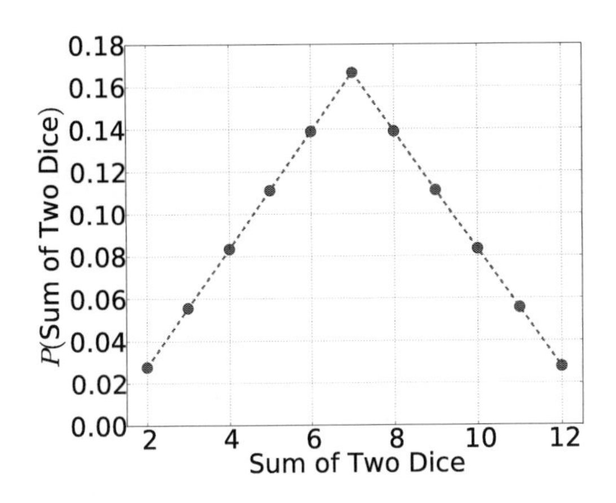

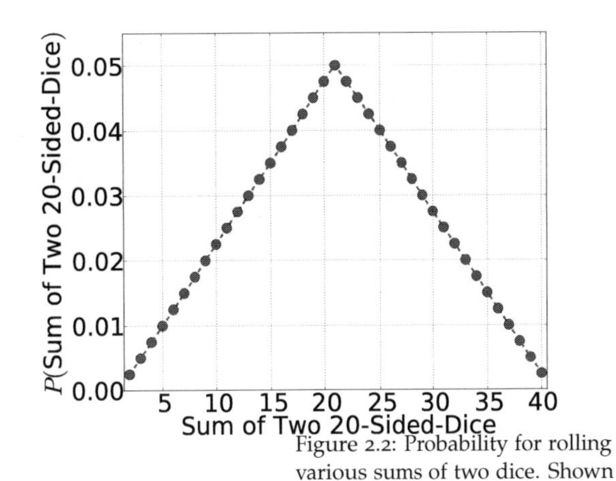

Figure 2.2: Probability for rolling various sums of two dice. Shown are the results for two 6-sided dice (left) and two 20-sided dice (right). The dashed line is for clarity, but represents the fact that you can't roll a fractional sum, such as 2.5.

10000 people who have been tested for cancer, and we get the following hypothetical data:

Number of Individuals	Negative Test	Positive Test	Total
Doesn't Have Cancer	9200	700	9900
Has Cancer	20	80	100
	9220	780	10000

We may be interested in a number of related probabilities.

EXAMPLE 2.16 *What is the probability of both having cancer and getting a positive test for it?*

We can determine this by simply dividing the person counts from the table

$$P\,(\text{cancer } \textbf{and} \text{ positive test}) = \frac{\text{\# of people with both cancer and positive test}}{\text{total \# of people}}$$
$$= \frac{80}{10000} = 0.008$$

Doing this process for every part of the table yields a posterior probability table, giving the probability for every combination of variables (i.e. with cancer and positive test, without cancer and positive test, etc...)

Posterior Probability	Negative Test	Positive Test	Total
Doesn't Have Cancer	0.92	0.07	0.99
Has Cancer	0.002	0.008	0.01
	0.922	0.078	1.0

EXAMPLE 2.17 *What is the probability of both not having cancer and getting a positive test for it?*

Reading off of the table, we have

$$P\left(\text{no cancer } \textbf{and} \text{ positive test}\right) = 0.07$$

This question, it turns out, is a not very interesting question. The type of question that *actually* arises in life is the following,

EXAMPLE 2.18 *What is the probability of having cancer* given *a positive test for it?*

Here we can perform the calculation in a couple of different ways, to give the (unintuitive) result.

1 *Counting the individuals.*

$$\begin{aligned} P\left(\text{cancer}|\text{positive test}\right) &= \frac{\text{\# of people with both cancer and positive test}}{\text{\# of people with a positive test}} \\ &= \frac{80}{780} = 0.0103 \end{aligned}$$

This is such a low probability, even though the bulk of people with cancer actually test positive, and a small fraction of those that don't have cancer test positive. It is because there are many more people without cancer, so even a small fraction of those that test positive outweighs those with cancer.

2 *Applying Product Rule*

Using the Product Rule (Section 1.4 on page 36), we have

$$\begin{aligned} P\left(\text{cancer}|\text{positive test}\right) &= \frac{P\left(\text{cancer } \textbf{and} \text{ positive test}\right)}{P\left(\text{positive test}\right)} \\ &= \frac{0.008}{0.078} = 0.103 \end{aligned}$$

where we have used the sum of the *Positive Test* column for $P\left(\text{positive test}\right)$. This is simply a shortcut to the *marginalization* process (Section 1.4 on page 40) - determine the probability of an event by adding up all of the possible conditional situations,

$$\begin{aligned} P\left(\text{positive test}\right) &= P\left(\text{no cancer } \textbf{and} \text{ positive test}\right) + P\left(\text{cancer } \textbf{and} \text{ positive test}\right) \\ &= = 0.07 + 0.008 = 0.078 \end{aligned}$$

2.8 *Exercises*

EXERCISE 2.1 *What is the probability that at least 3 people have the same birthday in a group of 50?*

EXERCISE 2.2 *Examine the case of Monte Hall with 4 doors, the host opening one door with a goat, and leaving you with a choice of 3. Should you switch? Does it matter which of the other two you choose?*

EXERCISE 2.3 *What is the probability of rolling various sums from two 9-sided dice?*

EXERCISE 2.4 *What is the probability of rolling an odd sum with two dice?*

EXERCISE 2.5 *What is the probability of rolling more than 7 from two 20-sided dice?*

EXERCISE 2.6 *Given the table above, determine the following quantities, and describe what they* mean*:*

1 $P(C = 1, T = 0)$

2 $P(C = 0)$

3 $P(C = 0) + P(C = 1)$

2.9 *Probability and Juries*

A problem of independence

As said in the beginning of Chapter 1 (Introduction to Probability), in 1968 a jury found defendant Malcolm Ricardo Collins and his wife defendant Janet Louise Collins guilty of second degree robbery. The prosecutor focussed on the the distinctive features of the dependence, and assigned a probability to each as follows[6]:

[6] J. Sullivan. People v. Collins , 68 cal.2d 319, 1968. URL http://scocal.stanford.edu/opinion/people-v-collins-22583

1 Partly yellow automobile 1/10

2 Man with mustache 1/4

3 Girl with ponytail 1/10

4 Girl with blond hair 1/3

5 Negro man with beard 1/10

6 Interracial couple in car 1/1000

He then followed with the calculation applying the product rule (Section 1.4 on page 36), to find the probability that *all* these things could have been observed:

$$\frac{1}{10}\times\frac{1}{4}\times\frac{1}{10}\times\frac{1}{3}\times\frac{1}{10}\times\frac{1}{1000}=\frac{1}{12,000,000}$$

The initial conviction was overturned for two primary reasons, one legal and one mathematical. The legal argument was that the prosecution had not established that these initial probabilities were supported by the evidence. However, the really devastating part of the argument was mathematical. As you may recall, the product rule used in this way assumes the *independence* of the terms (Section 1.4 on pageSection 37).

EXAMPLE 2.19 *Beard and Mustache - An Examination of Independence*

For an example, the proper product rule for two of the terms above would look like:

$$\begin{aligned}&P\left(\text{Man with beard }\textbf{and}\text{ Man with mustache}\right)=\\&\qquad P\left(\text{Man with mustache}|\text{Man with beard}\right)P\left(\text{Man with beard}\right)\end{aligned}$$

What the prosecutor was assuming is that these two items were *independent*, from which it would follow that

$$\begin{aligned}&P\left(\text{Man with beard }\textbf{and}\text{ Man with mustache}\right)=\\&\qquad P\left(\text{Man with mustache}\right)P\left(\text{Man with beard}\right)=\frac{1}{40}=0.025\end{aligned}$$

However, with a very brief thought, we notice that this is equivalent to saying

> Knowing the man has a beard tells us *nothing* about the probability of him having a mustache!

Clearly, it is not nearly as common to have a beard with no mustache than with one, so knowing that the man had a beard would nearly certainly imply that he had a mustache or,

$$\begin{aligned}&P\left(\text{Man with beard }\textbf{and}\text{ Man with mustache}\right)=\\&\qquad \underbrace{P\left(\text{Man with mustache}|\text{Man with beard}\right)}_{\sim 1}P\left(\text{Man with beard}\right)\sim\frac{1}{10}\end{aligned}$$

and the probability calculated, just from these two terms, is much higher than the prosecutor was communicating.

Similar sorts of absurdities occur with other terms, like "blond hair" and "pony tail", as well as others. Finally, even if it was the case that this is a somewhat rare combination, given the number of people in Los Angeles, one might be able to calculate the probability that there is at least one more couple satisfying these characteristics. Just like the lottery problem (Section 2.2 on page 52), it becomes likely that there are more couples in the area like this, and thus the ruling was overturned.

Another problem with independence

Another problem brought up in the opening of Chapter 1 (Introduction to Probability) is the case of Sally Clark. Sally Clark was convicted in 1999 of the murder of her two young sons[7]. In the case, the statistical argument was

[7] Lord Justice Kay. R vs Sally Clark, April 2003. URL http://www.bailii.org/ew/cases/EWCA/Crim/2003/1020.html

> Professor Meadow was asked if a figure of 1 in 8,543 reflected the risk of there being a single SIDS within such a family. He agreed that it was. A table from the CESDI report was placed before the jury. He was then asked if the report calculated the risk of two infants dying of SIDS in that family by chance. His reply was: *"Yes, you have to multiply 1 in 8,543 times 1 in 8,543 and I think it gives that in the penultimate paragraph. It points out that it's approximately a chance of 1 in 73 million."*

What he was doing was equating the following in the product rule (Section 1.4 on page 1.4):

$$P\,(\text{second child dying of SIDS}|\textbf{first}\text{ child dying of SIDS}) = P\,(\text{second child dying of SIDS})$$

which is equivalent to saying

> Knowing that the child dies of a [not well understood] disease tells us *nothing* about the probability of the second child dying of the same [not well understood] disease.

Clearly this is ridiculous, because if there is a common source to the disease, the one death certainly increases the probability of the second. Such a common source could be something shared in the environmental, or perhaps a genetic disposition in the family for the disease.

Prosecutor's Fallacy

Both of the cases above are examples of what is called the prosecutor's fallacy. It occurs when someone assumes that the prior probability of an event is equal to the probability that the defendant is innocent. A simple example is that "if a perpetrator is known to have the same blood type as a defendant and 10% of the population share that blood type; then to argue on that basis alone that the probability of the defendant being guilty is 90% makes the prosecutors's fallacy, in a very simple form."[8]

[8] Wikipedia. Prosecutor's fallacy — Wikipedia, the free encyclopedia, 2014. URL http://en.wikipedia.org/wiki/Prosecutor's_fallacy

Essentially the prosecutor is ignoring the number of people who match the rare event. Also, although double-deaths by SIDS are rare, they are much more common than double-murders! One really has to look at

$$P(\text{innocence}|\text{evidence})$$

which is not the same as

$$P(\text{evidence})$$

2.10 Computer Examples

Coin Flips

```
from sie import *
```

Generate a small list of data...

```
data=randint(2,size=10)
print data
```

```
[1 0 0 1 0 0 0 1 0 0]
```

Generate a slightly larger list of data...

```
data=randint(2,size=30)
print data
```

```
[1 1 1 0 0 0 0 1 1 1 1 0 1 1 0 1 1 0 0 1 1 1 0 1 0 0 1 1 0 0]
```

```
data=randint(2,size=(2000,10))
data
```

```
array([[1, 0, 1, ..., 1, 0, 0],
       [1, 1, 1, ..., 0, 1, 0],
       [0, 0, 1, ..., 0, 0, 0],
       ...,
       [0, 0, 0, ..., 1, 1, 0],
       [0, 1, 0, ..., 0, 1, 1],
       [0, 1, 1, ..., 1, 0, 1]])
```

We have here a large collection of numbers (20000 of them!), organized in 2000 rows of 10 columns. We can sum all of the 20000 values, or we can sum across columns or across rows, depending on what we want.

```
sum(data)  # add up all of the 1's
```

```
9988
```

```
sum(data,axis=0)  # sum up all of the columns
```

```
array([1011, 1010, 1001, 1051, 1001, 1008,  962,  990,  976,  978])
```

```
sum(data,axis=1)  # sum up all of the rows
```

```
array([3, 7, 3, ..., 5, 4, 6])
```

Typically the hist command makes its own bins, which may not center on the actual count values. That's why we call countbins(N), to make bins centered on the counts.

```
N=sum(data,axis=1)  # number of heads in each of many flips
hist(N,countbins(10))
xlabel('Number of Heads')
ylabel('Number of Flips')
```

```
<matplotlib.text.Text at 0x10856e990>
```

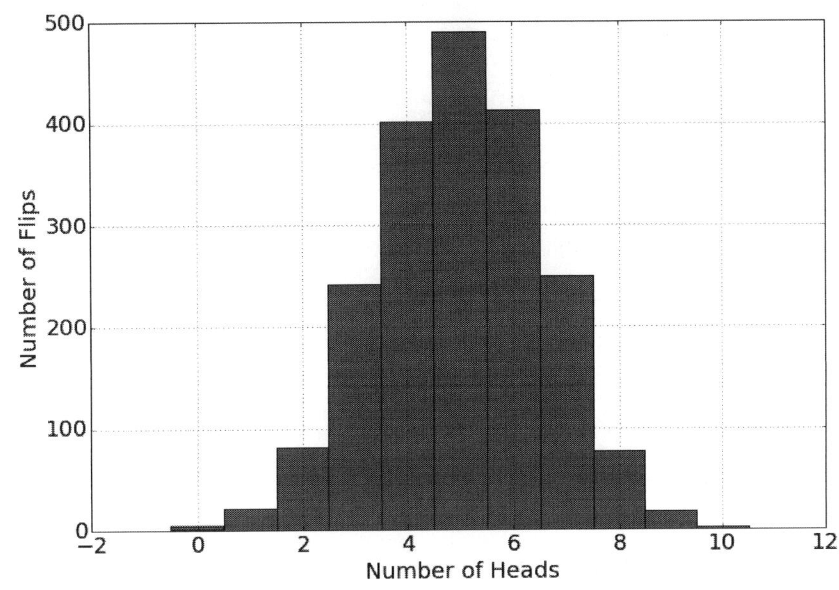

To get a probability distribution, we divide the histogram result by N.

This distribution is Bernoulli's equation, or in other words, the binomial distribution.

$$p(h,10) = \binom{10}{h} 0.5^h \cdot 0.5^{10-h}$$

```
h=array([0,1,2,3,4,5,6,7,8,9,10])

# or...

h=arange(0,11)
```

(recall that ** is exponentiation in Python, because the caret (^) was already used for a computer-sciency role.) The spaces in the equation below are not needed, but highlight the three parts of the binomial distribution.

```
p=nchoosek(10,h)* 0.5**h * 0.5**(10-h)
```

```
hist(N,countbins(10),normed=True)
plot(h,p,'--o')
xlabel('Number of Heads, $h$')
ylabel('$p(h|N=10)$')
```

```
<matplotlib.text.Text at 0x108560290>
```

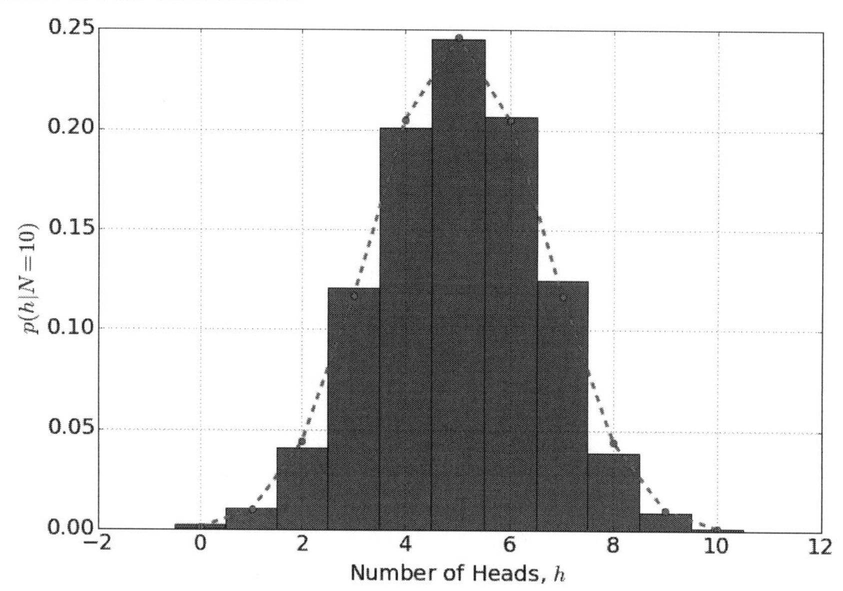

EXERCISE 2.7 *You flip a coin five times...*

1 *What is the probability of flipping 0, 1, 2, 3, 4, and 5 heads each in these 5 flips?*

2 *Show in a simulation that this matches these probabilities you just found.*

3 Random Sequences and Visualization

Now that we understand the rules of probability, and how they are applied in a number of practical examples, we explore the use of these rules to *sequences of random events*. This will produce several interesting and unintuitive observations and failures of inference, and the proper ways to handle them. Finally, we examine how visualize both data in general and what we can communicate with such visualization.

3.1 Coin Flipping

We'll start with some simple examples of coin flipping, asking some simple questions, and move to more complex observations and unintuitive conclusions.

EXAMPLE 3.1 *What is the probability of flipping three heads in a row, with a fair coin?*

We can approach this problem in two different ways. The first way, is a brute-force counting method with the definition of probability for exclusive events (using Equation 1.2) and the second way makes use of the other rules of probability. In the first way, we simply outline every possible combination of three flips, see how many are "three heads in a row"

All possible results from three coin flips:

1	T T T
2	T T H
3	T H T
4	T H H
5	H T T
6	H T H
7	H H T
8	H H H

Because there is only one case of "H H H" in all eight, the probability of three heads in a row is

$$P\,(\text{three heads in a row}) = 1/8$$

which is an unlikely outcome, but not extremely so (see Table 1.1).

In terms of the rule of probability, we have

$$P\,(\text{three heads in a row}) = P(H_1 \textbf{ and } H_2 \textbf{ and } H_3)$$

where H_1 is heads on the first flip, H_2 is heads on the second flip, etc... Because these are *independent events* (Section 1.4), the probability is just the product of the probabilities of the individual events

(Equation 1.9)

$$
\begin{aligned}
P\,(\text{three heads in a row}) &= P(H_1 \textbf{ and } H_2 \textbf{ and } H_3) \\
&= P(H_1) \times P(H_2) \times P(H_3) \\
&= \frac{1}{2} \times \frac{1}{2} \times \frac{1}{2} \\
&= \frac{1}{8}
\end{aligned}
$$

the same answer as before.

Yet again, we see that if there are multiple ways of arriving at an answer, that it must yield the same answer - equivalent states of knowledge yield equivalent probability assignments.

EXAMPLE 3.2 *What is the probability of flipping* thirty *heads in a row, with a fair coin?*

Our intuition will clearly insist that this will be a very small number, but how small? Our first method, of listing all of the possibilities gets quite a bit cumbersome with this question. The second method is quite straightforward

$$
\begin{aligned}
P\,(\text{thirty heads in a row}) &= P(H_1 \textbf{ and } H_2 \textbf{ and } \cdots \textbf{ and } H_{30}) \\
&= P(H_1) \times P(H_2) \times \cdots \times P(H_{30}) \\
&= \overbrace{\frac{1}{2} \times \frac{1}{2} \times \cdots \times \frac{1}{2}}^{\text{30 times}} \\
&= \left(\frac{1}{2}\right)^{30} \\
&= 0.000000001 \text{ (one in a billion!)}
\end{aligned}
$$

This is virtually impossible (Table 1.1).

EXAMPLE 3.3 *What is the probability of flipping two heads in three flips, with a fair coin?*

Our intuition suggests that this should be a reasonably common occurrence. We address this problem in exactly the same two ways: first, by counting, the second with the rules of probability. In the first method, we observe from the table that there are three ways of getting two heads: "T H H," "H T H," and "H H T." Thus,

$$
P\,(\text{two heads in three flips}) = \frac{3}{8}
$$

In the second method we write

$$
\begin{aligned}
&P\,(\text{two heads in three flips}) = \\
&\quad P\left((T_1 \textbf{ and } H_2 \textbf{ and } H_3) \textbf{ or } (H_1 \textbf{ and } T_2 \textbf{ and } H_3) \textbf{ or } (H_1 \textbf{ and } H_2 \textbf{ and } T_3)\right)
\end{aligned}
$$

from which we can apply the sum rule for exclusive events (Equation 1.11) and, like before, the product rule for independent events (Equation 1.9),

$$
\begin{aligned}
&P\,(\text{two heads in three flips}) = \\
&\qquad P(T_1 \textbf{ and } H_2 \textbf{ and } H_3) + P(H_1 \textbf{ and } T_2 \textbf{ and } H_3) + P(H_1 \textbf{ and } H_2 \textbf{ and } T_3) \\
&= \left(\frac{1}{2} \times \frac{1}{2} \times \frac{1}{2}\right) + \left(\frac{1}{2} \times \frac{1}{2} \times \frac{1}{2}\right) + \left(\frac{1}{2} \times \frac{1}{2} \times \frac{1}{2}\right) \\
&= \frac{1}{8} + \frac{1}{8} + \frac{1}{8} = \frac{3}{8}
\end{aligned}
$$

which is about a 38% chance, slightly unlikely (Table 1.1).

EXAMPLE 3.4 *What is the probability of flipping ten heads in thirty flips, with a fair coin?*

Once the numbers start getting large, our intuition fails, and we can't list all the possibilities. In order to proceed, we need to develop a systematic way of approaching these sorts of problems. Essentially it comes down to two parts:

1 What is the probability of *one particular sequence* being considered?

2 How many ways can this *type of sequence* appear in the process described in the question?

Point 1 is asking, what is the probability of this particular sequence:

H H H H H H H H H H T T T T T T T T T T T T T T T T T T T T

or this sequence:

T T H T T T T H H H H T T T T T H T T H T T T T T T H H T H

Although it is unintuitive, mathematically both of these specific sequences have *exactly* the same probability: each head or tail has equal probability, is not related to the others, and there are the same number of them. So we have

$$
\begin{aligned}
&P\,(\text{HHHHHHHHHHTTTTTTTTTTTTTTTTTTTT}) = \\
&\qquad P\,(\text{TTHTTTTHHHHTTTTTHTTHTTTTTTHHTH}) \\
&= \left(\frac{1}{2}\right)^{30} \\
&= 0.000000001 \text{ (one in a billion!)}
\end{aligned}
$$

Every single specific length-thirty sequence of heads and tails has the same probability, one in a billion.

Point 2 is asking, how many sequences are there of thirty heads and tails where ten of them are heads? Another way of phrasing it is, given a sequence like:

H H H H H H H H H H T T T T T T T T T T T T T T T T T T T T

how many different ways can I rearrange this sequence and get a unique sequence?

Counting the Rearrangements

We are going to determine the answer to our question in small steps. First, we ask,

EXAMPLE 3.5 *How many ways can we rearrange the unique symbols A, B, C, and D?*

To make this intuitive, we set up four empty boxes and we imagine placing our symbols in the boxes, one at a time. How many choices do we have? For the first box, we have four choices. For each of these choices, we've removed one of the symbols, and one of the boxes. Thus, we are left with three remaining symbols for each choice, and three remaining boxes. For each of the original four choices, we now have three choices for the second box. This immediately leads to $4 \times 3 = 12$ possibilities by the time we've filled two boxes. For each of these twelve possibilities, there are two symbols remaining and two boxes. Continuing this logic, we have two choices for the third box, and then only one choice for the final box. In summary, for each of the four choices for the first box we have three choices for the second, two choices for the third, and one for the final box. Thus we have

$$\begin{pmatrix}\text{number of rearrange-}\\ \text{ments of four different}\\ \text{symbols}\end{pmatrix} = 4 \times 3 \times 2 \times 1 = 24$$

Symbols: A B C D
Boxes: □□□□

Choices	Remaining Symbols
A□□□	B C D
B□□□	A C D
C□□□	A B D
D□□□	A B C

Choices	Remaining Symbols
A B□□	C D
A C□□	B D
A D□□	B C
B A□□	C D
B C□□	A D
B D□□	A C
C A□□	B D
C B□□	A D
C D□□	A B
D A□□	B C
D B□□	A C
D C□□	A B

In general we have

Number of Rearrangements of N Unique Symbols

$$\begin{aligned} C(N) &= N \times (N-1) \times \cdots \times 2 \times 1 \\ &= N! \end{aligned} \quad (3.1)$$

where we've introduced the notation for the *factorial of N* as $N!$.

Number of Rearrangements of N Unique Symbols

$$\begin{aligned} C(N) &= N \times (N-1) \times \cdots \times 2 \times 1 \\ &= N! \end{aligned}$$

EXAMPLE 3.6 *How many ways can we rearrange the symbols A, A, A, and D?*

By eye we can see that there are only four rearrangements of these symbols. How is this different from the previous question with four symbols? We can imagine going from the first question, with four unique symbols "A B C D," and replace both "B" and "C" with "A" to get it. "BC" and "CB" are different sequences of unique symbols. However, if we replace "B" with an "A" and "C" with an "A", both sequences become the same sequence, namely "AA". If we try to blindly apply Equation 3.1, the one for the number of rearrangements of *unique* symbols, to the case where there are duplicates, we will *overestimate* the number of rearrangements - we are over counting duplicate subsequences. Further, we can be specific about how much

Symbols: A A A D

Rearrangements
D A A A
A D A A
A A D A
A A A D

we are over counting and thus find a new equation which includes the possibility of duplicates.

For example, if we have three duplicates in a sequence, the number of over countings will be the number of possible rearrangements of three unique symbols, because all of these rearrangements result in the same sequence of duplicate symbols. Thus, our procedure should be,

$$\begin{aligned}\begin{pmatrix}\text{number of rear-}\\\text{rangements of "A}\\\text{A A D"}\end{pmatrix} &= \frac{\begin{pmatrix}\text{number of rearrange-}\\\text{ments of four } \textit{unique}\\\text{symbols}\end{pmatrix}}{\begin{pmatrix}\text{number of rearrange-}\\\text{ments of the over-}\\\text{counted duplicate three}\\\text{symbols}\end{pmatrix}}\\ &= \frac{4!}{3!}\\ &= \frac{4\times 3\times 2\times 1}{3\times 2\times 1}\\ &= 4\end{aligned}$$

Example 3.7 *How many ways are there of rearranging the symbols "A A A D D"?*

Following the same logic, we have

$$\overbrace{\underbrace{\text{A A A}}_{\substack{\text{3! ways of}\\\text{rearranging}\\\text{3 duplicates}}}\quad\underbrace{\text{D D}}_{\substack{\text{2! ways of}\\\text{rearranging}\\\text{2 duplicates}}}}^{\substack{\text{5! ways of}\\\text{rearranging}\\\text{5 } \textit{unique}\\\text{symbols}}}$$

$$\begin{aligned}\begin{pmatrix}\text{number of rear-}\\\text{rangements of}\\\text{"A A A D D"}\end{pmatrix} &= \frac{5!}{3!2!}\\ &= \frac{5\times 4\times 3\times 2\times 1}{(3\times 2\times 1)\times(2\times 1)}\\ &= \frac{120}{6\times 2}\\ &= 10\end{aligned}$$

All possible results of rearranging the symbols "A A A D D":

1	A A D D A
2	D A A D A
3	A D A D A
4	D A A A D
5	D A D A A
6	A A D A D
7	D D A A A
8	A D D A A
9	A A A D D
10	A D A A D

Sequences of Heads and Tails

Now we can return to our original question,

EXAMPLE 3.8 *What is the probability of flipping ten heads in thirty flips, with a fair coin?*

We broke it down into two parts:

$$P(h = 10, N = 30) = P\begin{pmatrix}\text{one sequence of} \\ \text{10 heads and 20} \\ \text{tails}\end{pmatrix} \times \begin{pmatrix}\text{number of re-} \\ \text{arrangements} \\ \text{of a length-30} \\ \text{sequence with} \\ \text{10 ``H'' and 20} \\ \text{``T''}\end{pmatrix}$$

1 What is the probability of *one particular sequence* being considered?

$$\begin{aligned} P\begin{pmatrix}\text{one sequence of} \\ \text{10 heads and 20} \\ \text{tails}\end{pmatrix} &= \left(\frac{1}{2}\right)^{10} \times \left(\frac{1}{2}\right)^{20} \\ &= \left(\frac{1}{2}\right)^{30} \\ &= 0.00000000093 \text{ (one in a billion!)} \end{aligned}$$

2 How many ways can this *type of sequence* appear in the process described in the question?

Because we have a length-thirty sequence of "H" and "T" with 10 duplicate "H" symbols and 20 duplicate "T," we have the following number of ways that this could occur (i.e. the number of rearrangements of these sequences):

$$\begin{aligned} \begin{pmatrix}\text{number of re-} \\ \text{arrangements} \\ \text{of a length-30} \\ \text{sequence with} \\ \text{10 ``H'' and 20} \\ \text{``T''}\end{pmatrix} &= \frac{30!}{10!20!} \\ &= 30045015 \end{aligned}$$

So the probability of flipping 10 heads in 30 flips is

$$\begin{aligned} P(h = 10, N = 30) &= \frac{30!}{10!20!}\left(\frac{1}{2}\right)^{30} \\ &= 30045015 \times 0.00000000093 \\ &= 0.028 \end{aligned}$$

which is extremely unlikely (Table 1.1).

In general we have

Probability of flipping h heads and t tails Given the probability of flipping a single heads as 1/2, and the total number of flips is $N = h + t$, we have the following equivalent forms:

$$P(h,t) = \frac{(h+t)!}{h!t!} \times \left(\frac{1}{2}\right)^h \times \left(\frac{1}{2}\right)^t \tag{3.2}$$

$$P(h,N) = \frac{N!}{h!(N-h)!} \times \left(\frac{1}{2}\right)^h \times \left(\frac{1}{2}\right)^{N-h}$$

$$P(h,N) = \binom{N}{h} \times \left(\frac{1}{2}\right)^h \times \left(\frac{1}{2}\right)^{N-h}$$

where we have introduced the notation that is sometimes used, called *choose*, read as "N choose h,"

$$\binom{N}{h} \equiv \frac{N!}{h!(N-h)!}$$

Probability of flipping h heads and t tails Given the probability of flipping a single heads as 1/2, and the total number of flips is $N = h + t$, we have the following probability for h heads and t tails:

$$P(h,t) = \frac{(h+t)!}{h!t!} \times \left(\frac{1}{2}\right)^h \times \left(\frac{1}{2}\right)^t$$

Shown in Figure 3.1 is the probability of flipping h heads in 30 flips, for each value of h from $h = 0$ (no heads or, in other words, 30 tails) up to $h = 30$ (all 30 heads). Clearly the most likely value is 15, but all of the numbers from 12 up to 18 have significant probability.

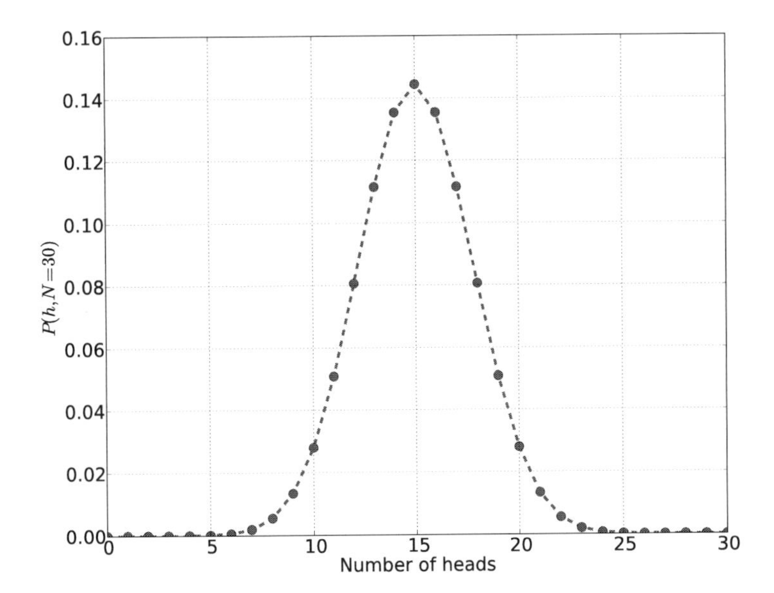

Figure 3.1: Probability of getting h heads in 30 flips. Clearly the most likely value is 15, but all of the numbers from 12 up to 18 have significant probability.

EXAMPLE 3.9 *What is the probability of getting 17 or more heads in 30 flips?*

Because these are independent events, we can simply sum up the terms for $P(h = 17, N = 30)$, $P(h = 18, N = 30)$, etc... yielding the following, either through direct calculation, or by reading the Figure 3.1.

$$
\begin{aligned}
P(h \geq 17, N = 30) &= \underbrace{0.11}_{h=17} + \underbrace{0.08}_{h=18} + \underbrace{0.05}_{h=19} + \underbrace{0.028}_{h=20} + \underbrace{0.013}_{h=21} + \underbrace{0.005}_{h=22} + \underbrace{0.002}_{h=23} + \underbrace{\text{(tiny numbers)}}_{h=23,24,25,26,27,28,29,30} \\
&= 0.29
\end{aligned}
$$

which is quite likely!

3.2 *Binomial Distribution*

The distribution of the possible number of *heads*, given N flips with a coin with probability p of flipping heads, is referred to as the Binomial Distribution. It has the form of Equation 3.3, with the "fair coin" probability, $1/2$, replaced with p:

$$
P(h|N,p) = \frac{N!}{h!(N-h)!} \times p^h \times (1-p)^{N-h} \tag{3.3}
$$

Probability of flipping h heads and t tails with an *unfair* coin Given the probability of flipping a single heads is, say, p and the total number of flips is $N = h + t$, we have the following equivalent forms:

$$
\begin{aligned}
P(h,t) &= \frac{(h+t)!}{h!t!} \times p^h \times (1-p)^t \qquad (3.4) \\
P(h,N) &= \frac{N!}{h!(N-h)!} \times p^h \times (1-p)^{N-h} \\
P(h,N) &= \binom{N}{h} \times p^h \times (1-p)^{N-h}
\end{aligned}
$$

where the probability of tails is $1 - p$.

Probability of flipping h heads and t tails with an *unfair* coin Given the probability of flipping a single heads as p, and the total number of flips is $N = h + t$, we have the following probability for h heads and t tails:

$$
P(h,t) = \frac{(h+t)!}{h!t!} \times p^h \times (1-p)^t
$$

3.3 *Streaks*

In the previous section we looked at the probability of getting a certain number of heads in a number of flips. Look at the following two sequences:

1 HTTHTHHTTHTHTTHHHTHHTTHHTHHTTHTHHTHHTTHTTHHHTHTHTT

2 HHTHHHTTTTTTTHTHTTHTTTHTHTHHTHTTHTTTHHTTTHHHHTHHHH

One of these sequences was generated from actually flipping a coin 50 times. The other one is from a person *pretending* to flip a coin, and writing down a sequence that they thought would look like a random flipping of a coin. Which one is which? While many people think

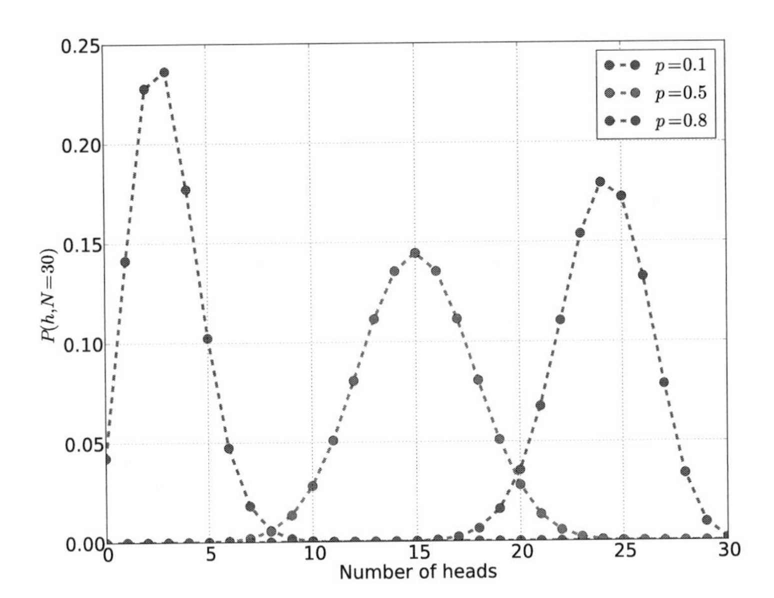

Figure 3.2: Probability of getting h heads in 30 flips given a possible unfair coin. One coin has $p = 0.1$, where the maximum is for 3 heads (or 1/10 of the 30 flips), but 2 heads is nearly as likely. Another has $p = 0.5$, and is the fair coin considered earlier with a maximum at 15 heads (or 1/2 of the 30 flips). Finally, another coin shown as $p = 0.8$ where 24 heads (or 8/10 of the 30 flips) is maximum.

that sequence 1 looks more "random" (i.e. it seems to flip around a lot), sequence 2 is actually the random sequence.

One of the truly unintuitive things about real random sequences, as opposed to designed sequences, is that there are long runs or *streaks*. Why is this? The general solution is beyond this book but we can think about it this way. Although a sequence of, say, 5 heads in a row is very unlikely (P (5 heads in a row) $= (1/2)^5 = 0.03$), there are many opportunities for such a sequence *somewhere* within a sequence of 50. Because of these many opportunities, this raises the probability from 3% (the probability of 5 heads in a row in 5 flips), to over 55%, the probability of finding 5 heads in a row *somewhere* in 50 flips. Streaks of 6 heads in a row occur nearly one third of the time in 50 flips, or over half the time if you consider a run to be either heads or tails. Even streaks of 9 heads or tails in a row, in 50 flips, is not extremely unlikely!

3.4 *Gambler's Fallacy*

When we look at a sequence of real coin flips, like:

- HHTHHHTTTTTTT

and we ask about the probability of flipping heads in the next flip, it is common to (mistakenly!) reason that, because we've seen 7 tails in a row, then the *next* flip is more likely to be heads. However, this is not the case for two reasons:

1. long streaks are common in completely fair and random sequences - so observing a streak of 7 tails does not contribute much to one's confidence that we are looking at a rigged coin or one that has changed its probability properties.

2. the process of flipping a coin is *independent* each time, nearly by definition, and thus the result of one flip cannot influence the result of the next flip.[1]

[1] One can imagine a flipping procedure where the flips are not independent. Say, you always place the resulting face (heads or tails) *initially up* in a flip, and the you do not flip particularly vigorously. Thus, the result of one flip would be related to the result of the next flip. However, in nearly all real cases, people go to great lengths to avoid this sort of procedure.

The faulty, but intuitive, reasoning goes by the name of the Gambler's Fallacy and appears in many places. We can ask a question:

> How could we tell the difference between a random, independent sequence and one where the events are not independent, where the next flip depended on a previous flip?

We'll have to return to this question later, when we consider model comparison, but roughly, one would have to look at all *pairs* of events to see if one pair (say heads-tails) occurs more frequently (even if only by a little) than another pair (say heads-heads).

In a total fit of irony, casino slot machines *do not produce independent winnings* - they are programmed so that if you've lost many times, then that machine is a little less likely to lose the next time. In effect, at gambling houses they train the gamblers in the Gambler's Fallacy!

3.5 The Hot Hand - Correlations in Random Sequences

Some work by Tversky and Gilovich[2] looks at the following issue in the sport of basketball: there are times when it seems as if basketball players have a "hot hand" - they are on a shooting streak. Tversky and Gilovich looked at how basketball fans *perceived* streaks, by having them rate sequences of shots as *random shooting* or *streak shooting*. Most (65%) of the respondents classified artificially generated, purely random sequences as *streak shooting*. In real data, they discovered that the actual probability of "making a given shot (i.e. a player's shooting *percentage*) is unaffected by the player's prior performance." We examine this effect in a later section (see Example 8.11 on page 158) where we explore the quantitative procedure for assessing this conclusion. It is enough here to note the large difference between the *perception* of the sequence and the likely *cause* of the sequence, and thus the need to always be vigilant against faulty perceptions. Tversky and Gilovich insist that "their observations do not tell us anything general about sports, but it does suggest a generalization about people, namely that they tend to 'detect' patterns even where none exist."

[2] A. Tversky and T. Gilovich. The cold facts about the" hot hand" in basketball. *Anthology of statistics in sports*, 16:169, 2005

What we have here, again, is the general perception that *long sequences* are somehow not "random," when in fact the opposite is the case. People have a natural tendency to see patterns in random data, to infer order where there is none, and to ascribe importance to the appearance of pattern. It is the role of statistical inference in general to provide the tools to properly handle the distinction between random effects and patterns, and to retune our intuitions.

3.6 Regression Toward the Mean

There is a peculiar phenomenon referred to as *regression toward the mean*, which often is misinterpreted and leads to failures of proper statistical inference. It can be seen in a simple example. Imagine that we "test" a number of students by having them guess the results of a coin flip. Clearly this will be entirely luck, because the coin flip has no pattern. If a student guesses the results of 50 flips, there will be an expectation of getting 25 correct. Here we simulate 20 students each "predicting" the result of 50 flips, the results shown in Table 3.1. The test is done twice, and we will look at a particular subset presently. One can, by eye, see that most of the students get around 25 correct - exactly as expected from random performance.

Now, imagine that we look at the *top five* coin flip predictors on the first round. Will they do better or worse in the the second round? What about the *bottom five* coin flip predictors? The results of these two cases are summarized in Table 3.2. The pattern, even in this small sample, is quite clear:

1 Those that did the best the first time did worse the second (on average)

2 Those that did the worst the first time did better the second (on average)

One might be tempted (had you not known that this is artificial data, and completely random) to interpret this as a causal pattern, e.g. "the students that did better the first time, grew over-confident the second," "the students that did worse the first time, worked harder to improve the second," etc... This *interpretation* of the results by students has been observed in the classroom.[3] However, it runs into serious trouble when the data is something like the heights of children compared to their parents - the tallest parents tend to have children shorter than they are, the shortest parents tend to have children taller than they are, a pattern first quantified by Galton in 1869[4]. He noted that clearly the children are not *trying* to be tall, so effort is not a good explanation for the pattern.

[3] Andrew Gelman and Deborah Nolan. *Teaching Statistics: A Bag of Tricks*. Oxford University Press, USA, 2002. ISBN 0198572247. URL http://www.amazon.com/Teaching-Statistics-A-Bag-Tricks/dp/0198572247

[4] F. Galton. *Hereditary Genius: An Inquiry Into Its Laws and Consequences*. Macmillan and Company, limited, 1914. URL http://books.google.com/books?id=bJB9AAAAMAAJ

Student	Total Correct First Round	Total Correct Second Round
1	23	24
2	23	29
3	19	23
4	26	27
5	28	29
6	26	22
7	23	26
8	30	28
9	24	21
10	27	23
11	25	31
12	30	21
13	20	22
14	28	29
15	24	25
16	25	22
17	23	24
18	20	28
19	20	29
20	28	25

Table 3.1: Total Correct Guesses from Students "Predicting" the Results of 50 Coin Flips. Shown are the results of a first round and a second round of guessing.

Top Five the First Time		**Bottom Five the First Time**	
Student	Performance the Second Time	Student	Performance the Second Time
8	Worse	3	Better
12	Worse	13	Better
14	Better	18	Better
5	Better	19	Better
20	Worse	1	Better

Table 3.2: Performance in the Second Round of Students "Predicting" the Results of 50 Coin Flips. Shown are the results for those students who performed *best* in the first round (left), and those that performed *worst* in the first round (right).

What is happening here is that, if the process is dominated by *luck* or simple random variation, then outliers occur, but are rare. Thus a particularly high value will likely be followed by a lower value - closer to the mean. The tendency is to regress *toward the mean* in processes dominated by luck. This can be confused with the Gambler's Fallacy discussed earlier, where flipping 3 heads in a row doesn't give you any information about flipping another heads - it is *not* more likely to be tails. Part of the difference is that we are dealing with a process that has *many possible values*, not just two, and thus we can have a mean value, and outliers.

When each of these ideas is applied to sports, the weather, or business there are some interesting conclusions.

1. even when the process is *entirely random*, long streaks occur - and are often misinterpreted as an increase in the probability of the event.

2. when a person performs very well at their job (a number of successful business decisions, a high batting average, etc...) they will often do *worse* the next year - and again many are surprised, and interpret the result as the person "losing their touch" - when in fact, they may just have been lucky for a bit, and are now performing closer to their typical average level.

3. when one has a particularly bad winter, it may be more likely that the next winter won't be quite do bad - due entirely to regression to the mean. It may, however, be part of a larger pattern (e.g. a large-scale climate oscillation, such as El Niño) and the probability of another bad winter might be *higher*. In order to tell the difference, we need to construct reasonable models of the phenomena, test those models with predictions, and apply those models into the future. At each step, we need to be careful not to jump to the conclusion of the existence of a pattern too quickly.

3.7 *Visualization of Data*

There are two main methods of visualizing data, and several others that are related to these methods. In this section we introduce just two, histograms and scatter plots, and we will use these throughout the text.

Histograms

Histograms are a way of *summarizing* data, when presenting the entire data set is impractical, or where some understanding of the data is made clearer by summarizing. The histogram plot is done with the following steps:

Another advantage to learning to understand how to generate histograms is that it alerts you to the possible *abuses* of these plots. These abuses can be simple mistakes, which end up giving a misleading message, or a deliberate deception. Either way, a proper understanding of the process helps.

1 Choose a number of *bins* to divide the data.

2 Count up the data that fall into each bin

3 Make a *bar plot*, or a *scatter plot* to present the data.

The following is an example with a small data set. The process of binning and counting is often done by computer, but it is instructive to perform the process by hand a few times in order to understand what the results are.

Table 3.3 shows a collection of 106 heights (in centimeters) of the male students in a class[5]. As a collection of numbers it is relatively opaque, but as a histogram it is clearer.

[5] Vincent Arel-Bundock. Rdatasets R datasets: An archive of datasets distributed with R, 2014. URL http://vincentarelbundock.github.io/Rdatasets/

177.8	160.0	165.0	182.88	175.0	167.0
182.88	190.5	177.0	190.5	180.34	180.34
184.0	172.72	175.26	167.0	180.0	180.0
190.0	182.5	185.0	171.0	172.0	180.34
180.0	170.0	200.0	190.0	170.18	179.0
182.0	171.0	177.8	175.26	187.0	183.0
180.0	176.0	185.42	176.5	167.64	179.0
183.0	179.0	190.0	165.0	187.0	170.0
180.0	180.34	190.5	185.0	193.04	184.0
177.0	180.0	175.26	180.34	178.5	187.96
178.0	175.26	189.0	182.88	170.0	180.0
185.0	187.96	185.42	195.0	172.72	180.34
173.0	187.96	187.0	168.0	191.8	177.0
189.0	180.34	182.88	172.72	172.0	170.0
175.0	168.0	165.0	173.0	196.0	179.1
180.0	176.0	154.94	174.0	179.1	160.0
165.0	165.0	170.0	185.0	188.0	171.0
185.0	185.0	180.34	183.0		

Table 3.3: 106 Male Student Heights (in cm) from a Survey.

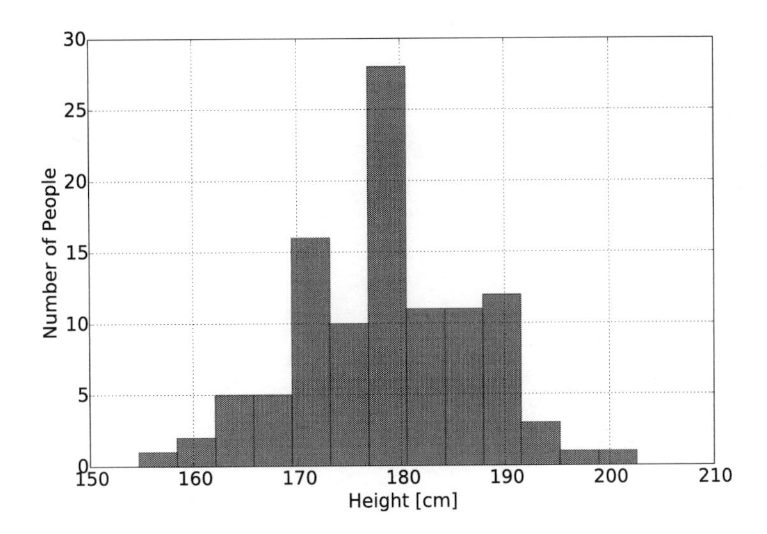

From this histogram, we can immediately observe several quantities which summarize there data:

1 The *average value* (around the middle) should be around 175 cm. The actual value can be calculated from the data, as

$$\bar{x} = \frac{177.8 + 160.0 + \cdots + 180.34 + 183.0}{106} = 178.83$$

2 The *range* of the data is around 155 cm up to about 205 cm. Again we can be more precise, and find the minimum of the data (154.94 cm) and the maximum (200 cm) but the histogram picture yields an approximate value instantly.

3 The values are *roughly symmetric* about the mean (i.e. average) value. This can give us a clue concerning how to model the data.

What is quite clear is that it is far easier to deal with a histogram, as above, than find the same information from the table of numbers.

Too Few Bins Plotting the same histogram with too few bins might look like:

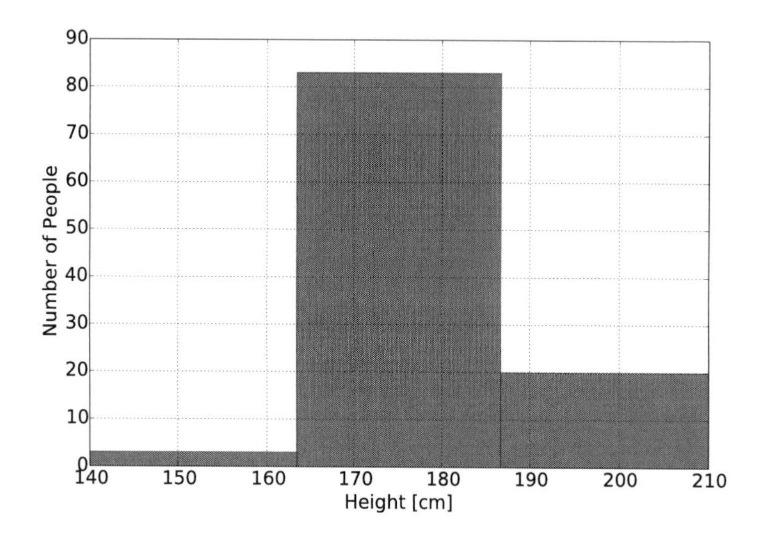

Clearly all the information is washed out.

Too Many Bins Plotting the same histogram with too many bins might look like:

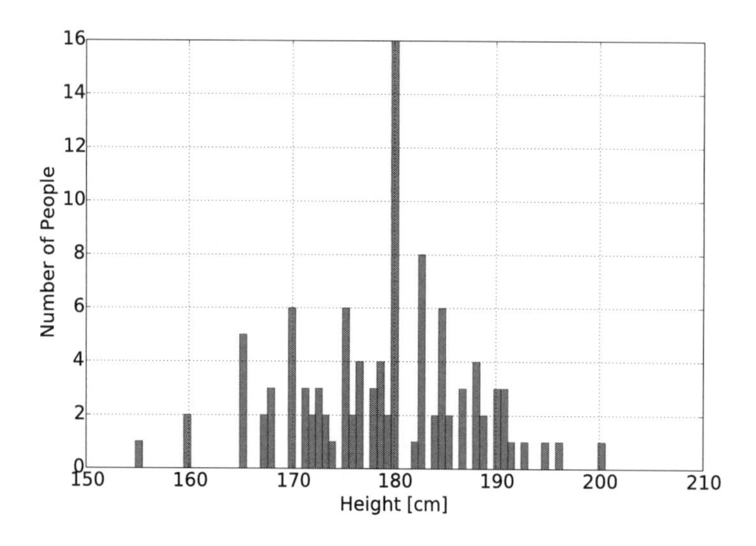

We lose any of the summary information here, where we essentially have one bar for each data-point.

Scatter Plots

A *scatter plot* is used to explore the relationship between two values. For example, in the survey of male students, in addition to height the students also measured the width of their writing hand viewed as a histogram, here

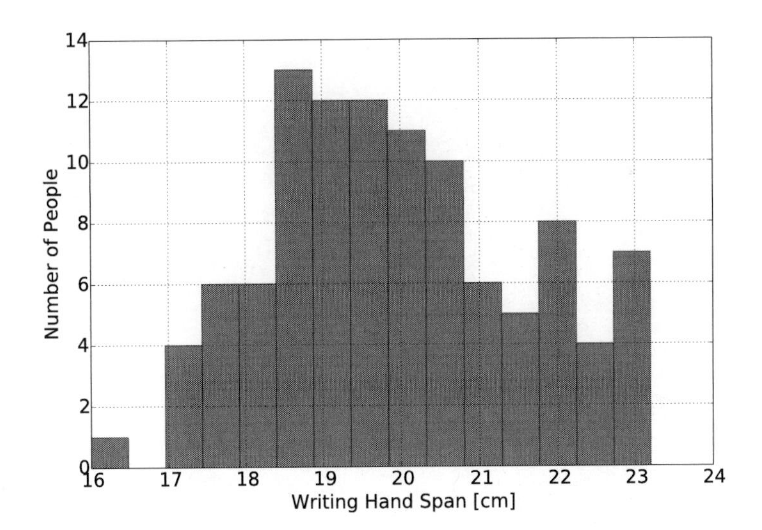

However, due to the possibility that these two variables could be *related*, it makes more sense to make a scatter plot. In such a plot, one designates one variable as "x" and another as "y," and places a *single dot* for each pair of values in the data set. Thus, each dot on the plot corresponds to height and hand-width for a *single* student.

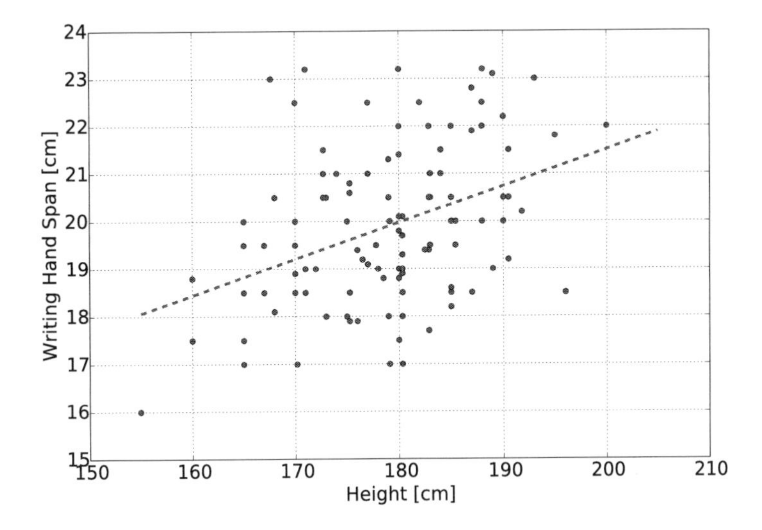

What we can see here, which was obscured with a histogram, is the *relationship* between these values - for the taller students, their hands are wider. We will explore quantifying this relationship later, but much can be done by eye using a scatter plot.

3.8 Computer Examples

This section summarizes how to make histograms and scatter plots with the computer software.

Histograms

```
from sie import *
```

Load a sample data set, and select only the Male data...

```
data=load_data('data/survey.csv')
male_data=data[data['Sex']=='Male']
```

select only the height data, and drop the missing data (na)...

```
male_height=male_data['Height'].dropna()
```

make the histogram

```
hist(male_height,bins=20)
xlabel('Height [cm]')
ylabel('Number of People')
```

```
<matplotlib.text.Text at 0x1085728d0>
```

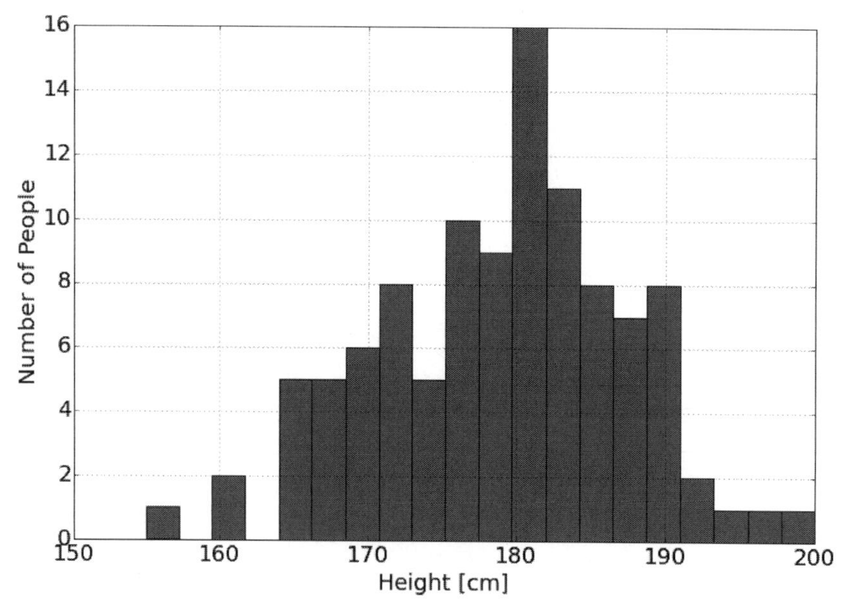

Scatter Plot

```
from sie import *
```

Load a sample data set, and select only the Male data...

```
data=load_data('data/survey.csv')
male_data=data[data['Sex']=='Male']
```

select only the height and the width of writing hand data, and drop the missing data (na)...

```
subdata=male_data[['Height','Wr.Hnd']].dropna()
height=subdata['Height']
wr_hand=subdata['Wr.Hnd']
```

plot the data

```
plot(height,wr_hand,'o')
ylabel('Writing Hand Span [cm]')
xlabel('Height [cm]')
```

```
<matplotlib.text.Text at 0x1085774d0>
```

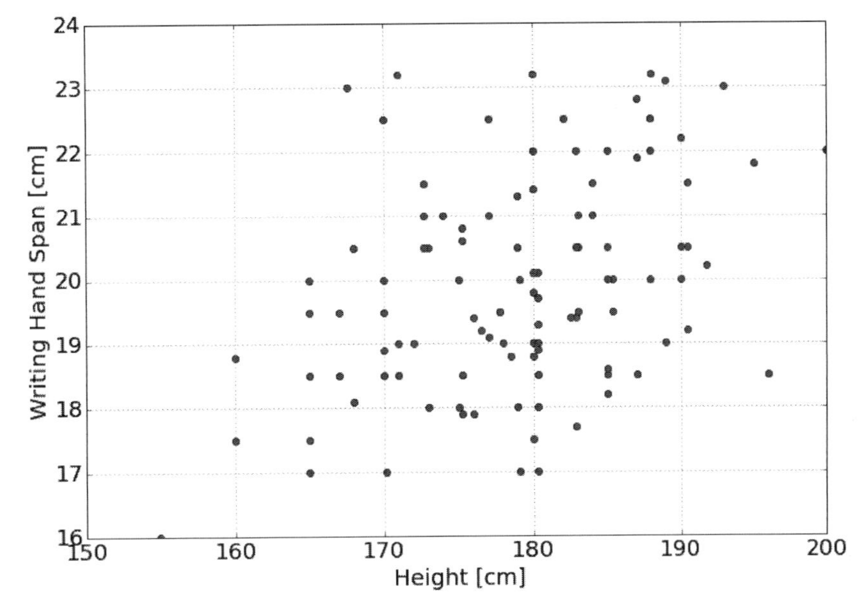

4 Introduction to Model Comparison

A *model*[1] as we use the term in this book is a *specific description of a possible state of nature*. This is in contrast to an *actual* state of nature, which we practically never have access to. We can never know *anything* with 100% certainty, and must therefore be open to alternate possible explanations, or models, describing our observations. For example, in medicine such models could include "I have lung cancer," "I have pneumonia," and "I have a cold." In physics, models could include "the Earth moves around the Sun" and "the Sun moves around the Earth." We can imagine many possible models that are consistent with the observed data, and our job in doing statistical inference is to determine the probabilities of our models given the data we observe. In our notation, what we are always looking for is

$$P(\text{model}|\text{data}) \tag{4.1}$$

We will explore model comparison through a series of examples.

[1] A similar term is *hypothesis*, and model comparison would then be *hypothesis testing*. We don't choose to use that term, partly because of the colloquial use of hypothesis as a kind of "guess," but also because hypothesis testing in some treatments focus on true/false tests of hypotheses which can lead to some significant misunderstandings. The use of models implies the possibility of multiple (i.e. more than two) models.

4.1 The High/Low Deck Game

In this example we use a simple card game as a platform for discussing model comparison in general. We start with two atypical decks of cards called the High Deck and the Low Deck (Figures 4.1 on page 90 and 4.2 on page 90 respectively). The game goes as follows.

> You're handed one of the two decks, but you don't know which. First, you draw the top card and note the value. Second, you replace the card and *reshuffle the deck*[2]. You repeat this procedure of drawing, noting, and reshuffling for as many turns as you need. The goal is to determine which of the the two decks (High or Low) you are in fact holding in your hand.

[2] Although we could make a game without replacement, which may be simpler to implement, the version of the game with reshuffling will help with an example later.

What does our intuition say?

We start by exploring our intuitions, before we do anything mathematically. Thus, we are in a position to check to see if the math is

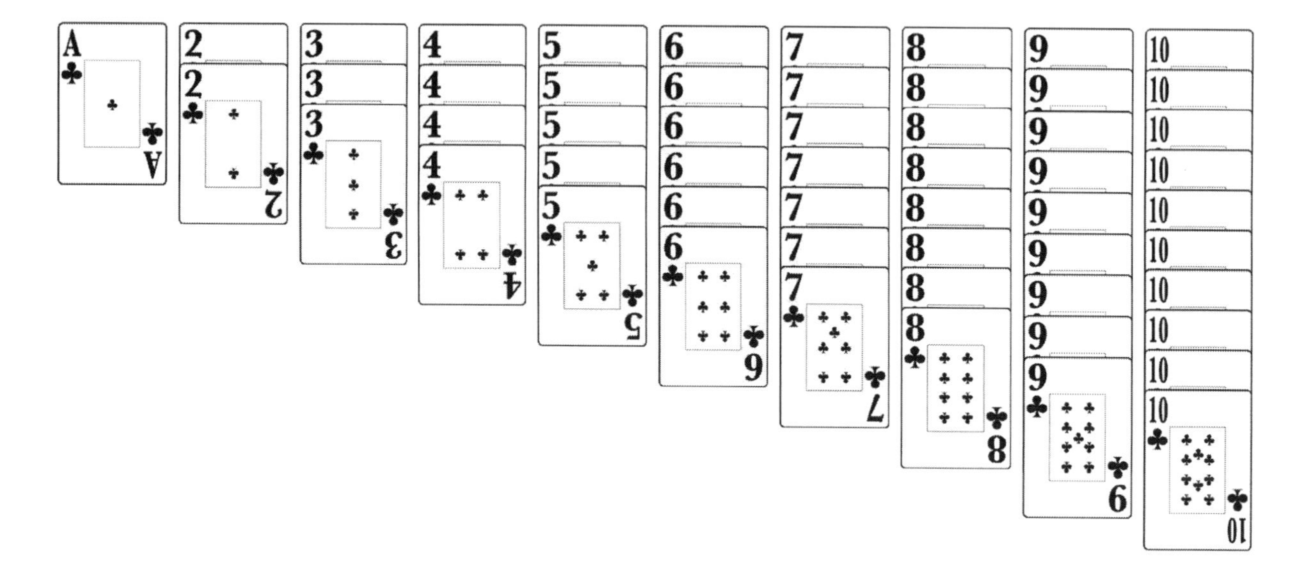

Figure 4.1: High Deck - 55 Cards with ten 10's, nine 9's, etc... down to one Ace. Aces are equivalent to the value 1.

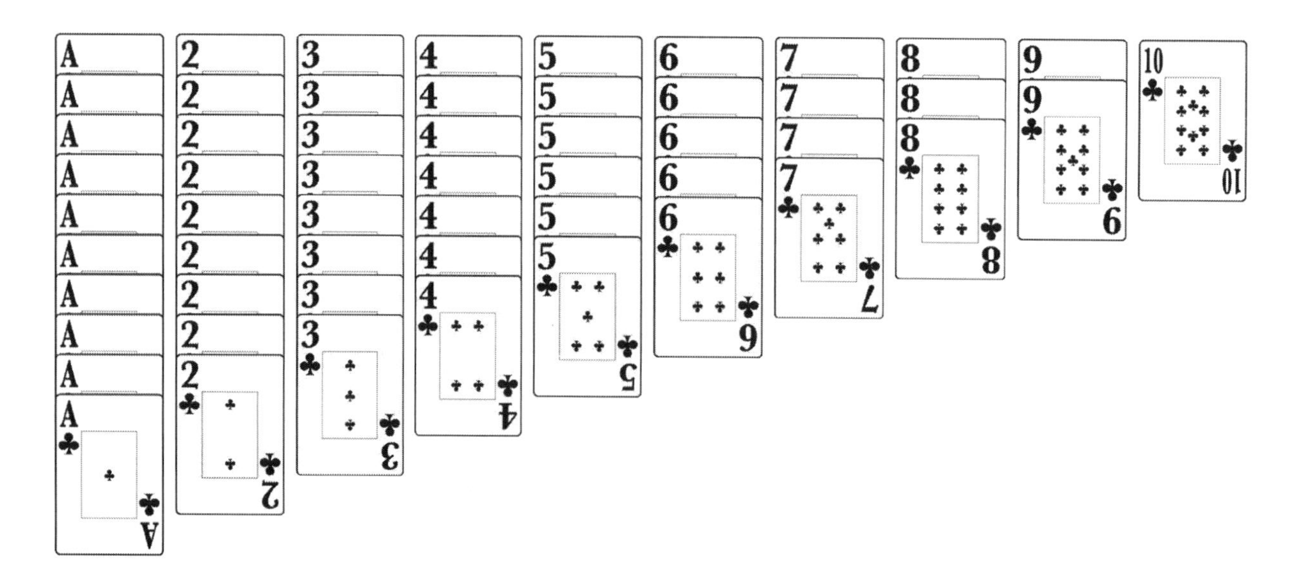

Figure 4.2: Low Deck - 55 Cards with ten Aces, two 2's, etc... up to one 10. Aces are equivalent to the value 1.

reasonable before we use the same math in areas where our intuition is not as strong. Imagine we draw only one card, and it is a 9. Intuition suggests that this constitutes reasonably strong evidence toward the belief that we're holding the High Deck. If we then (as the procedure states) place the 9 back in the deck, reshuffle and then draw a 7 we can be more strongly convinced that we are holding the High Deck. Repeating the reshuffle, and then drawing a 3 would make us a little less confident in this conclusion, but still quite certain. In this way we can sense how drawing different cards pushes our belief around, depending on how often that card comes up in the different decks.

Before the data - the prior

Before we take any data, we need to quantify our state of knowledge concerning all of the models that we are considering. In this case it is quite simple, because there are two models (High Deck and Low Deck), and we have been given no information about whether either is more common. With no such information, it is equivalent to a coin flip - we assign equal probabilities to both models *before* we see data, also known as the *prior* probabilities[3].

$$
\begin{aligned}
P(H) &= 0.5 \\
P(L) &= 0.5
\end{aligned}
$$

Surely this assessment will change *after* we see data, but that is the rest of the problem.

[3] The prior is sometimes mischaracterized as simply our guess, or some other completely subjective assessment of our knowledge. In fact in this example, and many others, we can make the *positive* case for equal probabilities given the state of our knowledge. This can be quantified with the concept of entropy, which is beyond this chapter.

The "easy" question - the likelihood

Although our ultimate goal is to infer the type of deck from the cards that we draw from it, we can start looking at an easier part of this question which serves as a first step toward the more challenging, and interesting goal. That question is the following,

EXAMPLE 4.1 *What is the probability of drawing a 9, given that we know that we're holding the High Deck?*

This related question is written

$$P(\text{data} = 9|H)$$

where data $= 9$ means that we have observed (i.e. drawn) one 9. This question is "easy" in the sense that it is simply related to the properties of the High Deck: the number of 9's and total number of cards. If you know that you have the high deck, then you know there are nine 9's in that deck out of 55 cards, and thus we have the probability of drawing one 9, given that we are holding the High Deck, is

$$P(\text{data} = 9|H) = \frac{9}{55}$$

We give this the name *likelihood*[4], and is simply the probability that the data could be the result of a known model. It is also the first part of the top of Bayes' Rule, Equation 1.14 on page 41.

[4] The term *likelihood* is a poorly chosen word. In English, this word is nearly synonymous with the word *probability* and thus easily leads to confusion. We could try to use a different term, like *consequent probability* or *generative likelihood* to stress the idea that the *likelihood* is the probability that the data we observe could be generated or could be a consequence of the particular model. However, we'd be going up against two centuries of continued use of the term *likelihood* and thus would probably increase confusion rather than decrease it.

Applying the Bayes' recipe

Here we introduce for the first time a recipe we will follow for all model comparison examples.

Now that we have our intuition, and we have the likelihoods, we can address the math. The two models are:

$$\begin{aligned} H &\equiv \text{"We're holding the High Deck"} \\ L &\equiv \text{"We're holding the Low Deck"} \end{aligned}$$

and the initial data is

$$\text{data} \equiv \text{"We've drawn one card, and it is a 9"}$$

According to Equation 4.1 on page 89 we are looking for the two probabilities:

$$\begin{aligned} &P(H|\text{data} = 9) \\ &P(L|\text{data} = 9) \end{aligned}$$

which are related to the *prior* and the *likelihood* via Bayes' Rule (Equation 1.14):

$$\begin{aligned} P(H|\text{data} = 9) &= \frac{P(\text{data} = 9|H)P(H)}{P(\text{data} = 9)} \\ P(L|\text{data} = 9) &= \frac{P(\text{data} = 9|L)P(L)}{P(\text{data} = 9)} \end{aligned}$$

To calculate actual numbers, we apply the Bayes' Recipe to this problem,

1 Specify the prior probabilities for the models being considered

$$\begin{aligned} P(H) &= 0.5 \\ P(L) &= 0.5 \end{aligned}$$

2 Write the top of Bayes' Rule for all models being considered

$$\begin{aligned} P(H|\text{data} = 9) &\sim P(\text{data} = 9|H)P(H) \\ P(L|\text{data} = 9) &\sim P(\text{data} = 9|L)P(L) \end{aligned}$$

where we are using the symbol $\sim$ to denote *proportionality* or *related to*. Essentially, by calculating the top of Bayes' Rule first, the numbers are not *equal* to the final (i.e. posterior) probabilities but must be rescaled to make sure that they add up to 1. This is done in the final step. Up until that rescaling, we use the symbol $\sim$ and think of it as *related to*.

3 Put in the likelihood and prior values

$$P(H|\text{data} = 9) \sim \frac{9}{55} \times 0.5 = 0.082$$
$$P(L|\text{data} = 9) \sim \frac{2}{55} \times 0.5 = 0.018$$

4 Add these values for all models

$$K = 0.082 + 0.018 = 0.1$$

5 Divide each of the values by this sum, K, to get the final probabilities

$$P(H|\text{data} = 0.082/0.1 = 0.82$$
$$P(L|\text{data} = 0.018/0.1 = 0.18$$

From which we can conclude that drawing a 9 does indeed constitute reasonably strong evidence toward the belief that we're holding the High Deck - the probability of us holding the High Deck, given the data, is 0.82.

Drawing the next card

So, when we draw a 7 next (after reshuffling), our intuition suggests that we'd be more confident that we're holding the High Deck. Repeating our recipe we have

The two models are:

$$H \equiv \text{"We're holding the High Deck"}$$
$$L \equiv \text{"We're holding the Low Deck"}$$

and data is

$$\text{data} \equiv \text{"We've drawn one card, and it is a 9, replaced and reshuffled, and then drawn a 7"}$$

According to Equation 4.1 we are looking for the two probabilities:

$$P(H|\text{data} = 9 \text{ then a } 7)$$
$$P(L|\text{data} = 9 \text{ then a } 7)$$

which are related to the *prior* and the *likelihood* via Bayes' Rule (Equation 1.14):

$$
\begin{aligned}
P(H|\text{data} = 9 \text{ then a } 7) &= \frac{P(\text{data} = 9 \text{ then a } 7|H)P(H)}{P(\text{data} = 9)} \\
P(L|\text{data} = 9 \text{ then a } 7) &= \frac{P(\text{data} = 9 \text{ then a } 7|L)P(L)}{P(\text{data} = 9)}
\end{aligned}
$$

To calculate actual numbers, we apply the Bayes' recipe to this problem,

1 Specify the prior probabilities for the models being considered

$$
\begin{aligned}
P(H) &= 0.5 \\
P(L) &= 0.5
\end{aligned}
$$

2 Write the top of Bayes' Rule for all models being considered

$$
\begin{aligned}
P(H|\text{data} = 9 \text{ then a } 7) &\sim P(\text{data} = 9 \text{ then a } 7|H)P(H) \\
P(L|\text{data} = 9 \text{ then a } 7) &\sim P(\text{data} = 9 \text{ then a } 7|L)P(L)
\end{aligned}
$$

3 Put in the likelihood and prior values

$$
\begin{aligned}
P(H|\text{data} = 9 \text{ then a } 7) &\sim \frac{9}{55} \times \frac{7}{55} \times 0.5 = 0.0104 \\
P(L|\text{data} = 9 \text{ then a } 7) &\sim \frac{2}{55} \times \frac{4}{55} \times 0.5 = 0.0013
\end{aligned}
$$

4 Add these values for all models

$$
K = 0.0104 + 0.0013 = 0.0117
$$

5 Divide each of the values by this sum, K, to get the final probabilities

$$
\begin{aligned}
P(H|\text{data} = 9 \text{ then a } 7) = 0.0104/0.0117 = 0.889 \\
P(L|\text{data} = 9 \text{ then a } 7) = 0.0013/0.0117 = 0.111
\end{aligned}
$$

which again matches our intuition - we're more confident that we're holding the High Deck, now with probability 0.889 increased from 0.82 when we just observed the 9.

Prior information or not?

In the above example, we started with a prior probability of holding the High Deck at $P(H) = 0.5$, because we had no information other than that there were two possibilities. We then observed a 9, and updated the probability to 0.82, and then observed a 7, and further updated the probability to 0.889 - making it more likely that we were holding the High Deck. One of the basic tenets of probability theory is that if there is more than one way to arrive at an answer, one should arrive at the same answer.[5] In the above, we calculated the probability of holding the High Deck given the observed data

[5] E. T. Jaynes uses the principle that "if there is more than one way to arrive at an answer, one should arrive at the same answer" to help derive the rules of probability from first principles. Failures of this principle result in paradoxes. This principle is also applied in Section 2.1 for the birthday problem.

$$\text{data} \equiv \text{"We've drawn one card, and it is a 9, replaced and reshuffled, and then drawn a 7"}$$

and prior information

$$\text{prior} \equiv \text{"We only know there are two decks."}$$

An equivalent situation is found after our first draw, after we've observed the 9, and we're about to draw our second card. In this case we have the *prior* information:

$$\text{prior} \equiv \text{"We only know there are two decks, and then we draw one card and it is a 9, replace it and reshuffle."}$$

and observed data:

$$\text{data} \equiv \text{"We've drawn one card and it is a 7"}$$

Mathematically, we apply the Bayes' recipe, but with the different prior information

1 Specify the prior probabilities for the models being considered

$$\begin{aligned} P(H,9) &= 0.82 \\ P(L,9) &= 0.18 \end{aligned}$$

2 Write the top of Bayes' Rule for all models being considered

$$\begin{aligned} P(H|\text{data} = 9 \text{ then a } 7) &\sim P(\text{data} = \{7\}|H)P(H,9) \\ P(L|\text{data} = 9 \text{ then a } 7) &\sim P(\text{data} = \{7\}|L)P(L,9) \end{aligned}$$

3 Put in the likelihood and prior values

$$\begin{aligned} P(H|\text{data} = 9 \text{ then a } 7) &\sim \frac{7}{55} \times 0.82 = 0.104 \\ P(L|\text{data} = 9 \text{ then a } 7) &\sim \frac{4}{55} \times 0.18 = 0.013 \end{aligned}$$

4 Add these values for all models

$$K = 0.104 + 0.013 = 0.117$$

5 Divide each of the values by this sum, K, to get the final probabilities

$$P(H|\text{data} = 9 \text{ then a } 7) = 0.104/0.117 = 0.889$$
$$P(L|\text{data} = 9 \text{ then a } 7) = 0.013/0.117 = 0.111$$

yielding the same result.

In other words our *updated probabilities* from the first draw can be seen as our prior probabilities for the subsequent draws. Thus, Bayes' Rule describes how we update our knowledge with new evidence, or in other words, *learning*.

4.2 Multiple Hypotheses

We start this section with an example.

EXAMPLE 4.2 *What is the probability they you are holding one of either the High or the Low Deck having drawn five 9's in a row from that deck?*

We have observed the following data:

$$\text{data} \equiv \begin{cases} \text{"We've drawn one card, and it is a 9, replaced} \\ \text{and reshuffled, redrawn and observed another} \\ \text{9, repeated this procedure and observed three} \\ \text{more 9's, for a total of five 9's in a row."} \end{cases}$$

Technically, drawing 5 9's in a row should give us really strong confidence that you are drawing from the High Deck, because we would have

1 Specify the prior probabilities for the models being considered

$$P(H) = 0.5$$
$$P(L) = 0.5$$

2 Write the top of Bayes' Rule for all models being considered

$$P(H|\text{data} = 5 \text{ 9's in a row}) \sim P(\text{data} = 5 \text{ 9's in a row}|H)P(H)$$
$$P(L|\text{data} = 5 \text{ 9's in a row}) \sim P(\text{data} = 5 \text{ 9's in a row}|L)P(L)$$

3 Put in the likelihood and prior values

$$P(H|\text{data} = 5 \text{ 9's in a row}) \sim \underbrace{\frac{9}{55} \times \frac{9}{55} \cdots \frac{9}{55}}_{5 \text{ times}} \times P(H)$$

$$
\begin{aligned}
& \sim \left(\frac{9}{55}\right)^5 \times 0.5 \\
& = 0.0000587 \\
P(L|\text{data} = 5 \text{ 9's in a row}) & \sim \left(\frac{2}{55}\right)^5 \times 0.5 \\
& = 0.0000000318
\end{aligned}
$$

4 Add these values for all models

$$K = 0.0000587 + 0.0000000318 = 0.0000587318$$

5 Divide each of the values by this sum, K, to get the final probabilities

$$
\begin{aligned}
P(H|\text{data} = 5 \text{ 9's in a row}) &= \frac{0.0000587}{0.0000587318} = 0.99946 \\
P(L|\text{data} = 5 \text{ 9's in a row}) &= \frac{0.0000000318}{0.0000587318} = 0.00054
\end{aligned}
$$

which is *fantastically* on the side of the high deck, even though we might start getting suspicious in this situation.

EXAMPLE 4.3 *What is the probability they you are holding one of either the High or the Low Deck having drawn m 9's in a row from that deck?*

In general, if we look at m 9's in a row, where m could be 1, 2, 3, etc..., we can see this following the Bayes' Recipe

1 Specify the prior probabilities for the models being considered

$$
\begin{aligned}
P(H) &= 0.5 \\
P(L) &= 0.5
\end{aligned}
$$

2 Write the top of Bayes' Rule for all models being considered

$$
\begin{aligned}
P(H|\text{data} = m \text{ 9's in a row}) &\sim P(\text{data} = m \text{ 9's in a row}|H)P(H) \\
P(L|\text{data} = m \text{ 9's in a row}) &\sim P(\text{data} = m \text{ 9's in a row}|L)P(L)
\end{aligned}
$$

3 Put in the likelihood and prior values

$$
\begin{aligned}
P(H|\text{data} = m \text{ 9's in a row}) &\sim \underbrace{\frac{9}{55} \times \frac{9}{55} \cdots \frac{9}{55}}_{m \text{ times}} \times P(H) \\
&\sim \left(\frac{9}{55}\right)^m \times 0.5 \\
P(L|\text{data} = m \text{ 9's in a row}) &\sim \left(\frac{2}{55}\right)^m \times 0.5
\end{aligned}
$$

4 Add these values for all models

$$K = \left(\frac{9}{55}\right)^m \times 0.5 + \left(\frac{2}{55}\right)^m \times 0.5$$

5 Divide each of the values by this sum, K, to get the final probabilities This step is easiest done in a table (Table 4.1), because the resulting expression is pretty messy.

m	$P(H\|\text{data})$	$P(L\|\text{data})$
1	0.81818	0.18182
2	0.95294	0.047059
3	0.98915	0.010855
4	0.99757	0.0024327
5	0.99946	0.00054163
6	0.99988	0.00012041
7	0.99997	0.000026761
8	0.99999	0.0000059470

Table 4.1: Drawing m 9's in a row, from either a High Deck or Low Deck.

It is clear from Table 4.1 that after drawing five 9's using our procedure, it should be *extraordinarily* likely that we are holding the High Deck. However, after a certain number of 9's observed, something starts to bother us. Perhaps not after five 9's, but what if the procedure were repeated and we drew ten 9's in a row? Or perhaps twenty 9's. At some point, we'd refuse to believe this is the High Deck because, although it was true that there are more 9's in the High Deck, there are many more *other* cards in the High Deck that we should see. What do we do in this case?

EXAMPLE 4.4 *What is the probability they you are holding one of either the High, Low, or Nines Deck having drawn m 9's in a row from that deck?*

The proper thing to do is to introduce a new model, say, a Nines deck. Clearly this model should have a very low prior probability, because we didn't even consider it before we saw the streak of 9's. Let's say that we assign the prior probability for the Nines deck to be a one in a million. To make all of the prior probabilities add up to 1, then the prior probabilities for the High and Low Deck must be a little less than 0.5. After that, we simply apply the Bayes' Recipe as before

What is interesting here is that once we admit that there are many possible models we could consider, we realize that we have these models in our head all the time, or we construct them as we need them. Every model comparison is a multiple model comparison, with most of the models with very low prior probabilities that our brain naturally suppresses until needed. Mathematically, we need to unsuppress them as needed.

1 Specify the prior probabilities for the models being considered

$$\begin{aligned} P(N) &= \frac{1}{1,000,000} = 0.000001 \\ P(H) &= 0.4999995 \\ P(L) &= 0.4999995 \end{aligned}$$

2 Write the top of Bayes' Rule for all models being considered

$$\begin{aligned} P(N|\text{data} = m \text{ 9's in a row}) &\sim P(\text{data} = m \text{ 9's in a row}|N)P(N) \\ P(H|\text{data} = m \text{ 9's in a row}) &\sim P(\text{data} = m \text{ 9's in a row}|H)P(H) \\ P(L|\text{data} = m \text{ 9's in a row}) &\sim P(\text{data} = m \text{ 9's in a row}|L)P(L) \end{aligned}$$

3 Put in the likelihood and prior values

$$\begin{aligned} P(N|\text{data} = m \text{ 9's in a row}) &\sim 1 \times P(N) = 0.000001 \\ P(H|\text{data} = m \text{ 9's in a row}) &\sim \underbrace{\frac{9}{55} \times \frac{9}{55} \cdots \frac{9}{55}}_{m \text{ times}} \times P(H) \\ &\sim \left(\frac{9}{55}\right)^m \times 0.4999995 \\ P(L|\text{data} = m \text{ 9's in a row}) &\sim \left(\frac{2}{55}\right)^m \times 0.0.4999995 \end{aligned}$$

4 Add these values for all models

$$K = 0.000001 + \left(\frac{9}{55}\right)^m \times 0.4999995 + \left(\frac{2}{55}\right)^m \times 0.4999995$$

5 Divide each of the values by this sum, K, to get the final probabilities Again, this step is easiest done in a table or, even better, a picture (Figure 4.3).

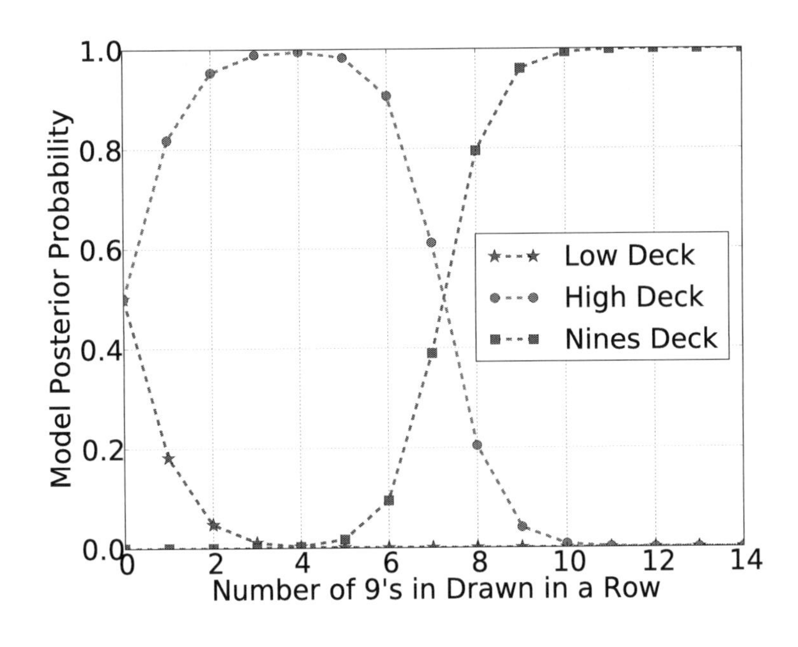

Figure 4.3: Drawing a number of 9's in a row, possibly from a High, Low, and Nines deck.

We have a clear picture here in Figure 4.3. As we initially draw 9's, our confidence that we're holding the High Deck goes up, at the expense of our confidence that we're holding the Low Deck. At a certain point (around six 9's in our example), our confidence in the High Deck starts to drop and we become more confident that something odd is happening, and our previously ignored model of the Nines deck becomes more likely. Eventually, this new model is the one in which we are the most confident.

Imagine further that if, after drawing ten 9's in a row we draw a 1. What do we do then? The likelihood for the Nines deck goes to *zero* instantly - the probability of drawing a 1 from a Nines deck is zero, $P(1|N) = 0$. Are we left again with the original two models, High and Low Deck? No! We would then introduce other models, perhaps something like a Mostly Nines Deck, or perhaps a High Deck with a weird shuffling procedure, or perhaps others. No matter how many models one has, the recipe is still the same. It is important to realize that in any model comparison case, there are always other models that could be brought to bear on the problem, perhaps with low prior probability. Simply showing that a model is consistent with a set of data does not insure against the possibility that another model could be better, if we could only think of it.

The creative part of science is not in the calculations performed, but in the generation of new and useful models. Until we come up with a better model for our data we make do with the ones that we have, all the while being aware that a better model may come into play later. Newton's Theory of Gravity was used for over 200 years, even when there was known data that made it less likely, until it was replaced by Einstein's Theory of Gravity. Newton's Laws, however, are still used in nearly all gravitational calculations because it is "good enough" and is a lot easier to work with practically.

EXERCISE 4.1 *Complete the example demonstrating the updated probabilities for the High and Low Deck, having drawn a 9, 7, and a 3. Compare with the case of drawing just the 9 and the 7, and discuss how it matches your intuition.*

EXERCISE 4.2 *Repeat the analysis of the sequence of 9's drawn in a row with an added hypothesis of a deck with one hundred 9's and one 8. Discuss the results. Demonstrate what happens to the probabilities for all of the hypotheses after drawing one 8, after ten 9's in a row. Discuss.*

EXERCISE 4.3 *I tell you that I have a coin that could have* both sides *heads,* both sides *tails, or a normal single-heads single-tails coin.*

1 *Before seeing the data, what would be a reasonable prior probability for the three hypotheses* H_0 *(no-heads),* H_1 *(one head), and* H_2 *(two heads)?*

2 *Would this have been different if you had simply been given a coin by a friend to flip to see who has to do the dishes? Why or why not?*

3 *Now I flip the coin once, and get a heads. Write down the* likelihood *of this data given each of the models. In other words, what are the values of:*

- $P\,(data{=}1\ heads|H_0)$
- $P\,(data{=}1\ heads|H_1)$
- $P\,(data{=}1\ heads|H_2)$

4 *Apply Bayes' Recipe, and determine the probability of each of these three models given this data. In other words, what are the values of:*

- $P(H_0|data{=}1\ heads)$
- $P(H_1|data{=}1\ heads)$
- $P(H_2|data{=}1\ heads)$

5 *Apply this recipe for the case of observing 3 heads in a row.*

5 *Applications of Model Comparison*

This chapter presents several applications of the model comparison concepts introduced in Chapter 4 (*Introduction to Model Comparison*).

5.1 *Disease Testing*

Let's imagine there is a rare, one in a million, disease that is lethal but does not have many outward symptoms at first. A new test boasts 99.9% accuracy, so you go to get tested, and receive the bad news that you test positive for the disease. Should you be devastated by the news? What is the probability that you *actually* have the disease? We are looking at two, quite different, probabilities here. In the first case, we have the claims of the test which state that *if you have the disease, the probability that the test will be positive is 0.999*, or, *if you have the disease, test will discover that fact 99.9% of the time*. In the second case we have your concern which is, *if you test positive for the test, what is the probability that you have the disease*. In our notation this is:

$$\begin{aligned} P(\text{positive test}|\text{disease}) &= 0.999 \text{ (claim from test)} \\ P(\text{disease}|\text{positive test}) &= \text{? (your concern)} \end{aligned}$$

These two are related by Bayes' Rule (Equation 1.14).
The Bayes' Recipe proceeds as follows

1 Specify the prior probabilities for the models being considered

The models we have are simply "have the disease" and "don't have the disease". The prior probabilities for these two come from the prevalence of the disease in the population, before you get tested. Since this is a "one in a million" disease, we have

$$\begin{aligned} P\left(\text{disease}\right) &= \frac{1}{1,000,000} \\ P\left(\text{no disease}\right) &= \frac{999,999}{1,000,000} \end{aligned}$$

2 Write the top of Bayes' Rule for all models being considered

The top of Bayes' Rule comes down to, given the truth of the model (i.e. either with or without the disease), what is the probability of getting the data (i.e. the positive or negative test result). This is measured by how good the test is.

In many medical applications, the false positive rate (P (positive test|no disease)) is not always equal to the false negative rate (P (negative test|disease)), so to say that a test is 99.9% accurate is actually incomplete - one needs to specify both rates of effectiveness. In this case, we are assuming that they are the same.

$$P\left(\text{positive test}|\text{disease}\right) = 0.999$$

and

$$P\left(\text{positive test}|\text{no disease}\right) = 0.001$$

So the top of Bayes' Rule looks for both models looks like:

$$\begin{aligned}
P\left(\text{disease}|\text{positive test}\right) &\sim P\left(\text{positive test}|\text{disease}\right) \times P\left(\text{disease}\right) \\
&\sim 0.999 \times \frac{1}{1,000,000} = 9.99 \times 10^{-7} \\
P\left(\text{no disease}|\text{positive test}\right) &\sim P\left(\text{positive test}|\text{no disease}\right) \times P\left(\text{no disease}\right) \\
&\sim 0.001 \times \frac{999,999}{1,000,000} = 9.99 \times 10^{-4}
\end{aligned}$$

3 Add these values for all models

$$K = 9.99 \times 10^{-7} + 9.99 \times 10^{-4} = 0.000999999$$

4 Divide each of the values by this sum, K, to get the final probabilities

$$\begin{aligned}
P\left(\text{disease}|\text{positive test}\right) &= \frac{9.99 \times 10^{-7}}{0.000999999} = 0.1\% \\
P\left(\text{no disease}|\text{positive test}\right) &= 99.9\%
\end{aligned}$$

Which means that, *overwhelmingly*, if you have a rare one-in-a-million disease, you are very unlikely to have it *even given a 99.9% accurate positive test for it*! This is a seriously unintuitive result, so it is helpful to visualize it in another way to build your intuition.

One way to see this result is to visualize it, as in Figure 5.1. Here, the numbers are a bit smaller - the disease is 1 out of 200 in a population of 3000, and the test is 99% accurate. This means about 15 sick people and about 2985 healthy people. If all of the sick people test positive, and 1% of the healthy people test positive due to the 99% accuracy, we would have 15 sick and 29 healthy people who all test positive. Even in this case, with much smaller numbers, we see that getting a positive test alone does not imply that it is likely you have the disease. It depends on the rarity of the disease (the more rare, the less likely) and the false positive rate (the number of healthy people who test positive anyway). This will vary depending on the disease and the test, but can lead to this unintuitive result, and thus can lead one to make poor medical decisions.

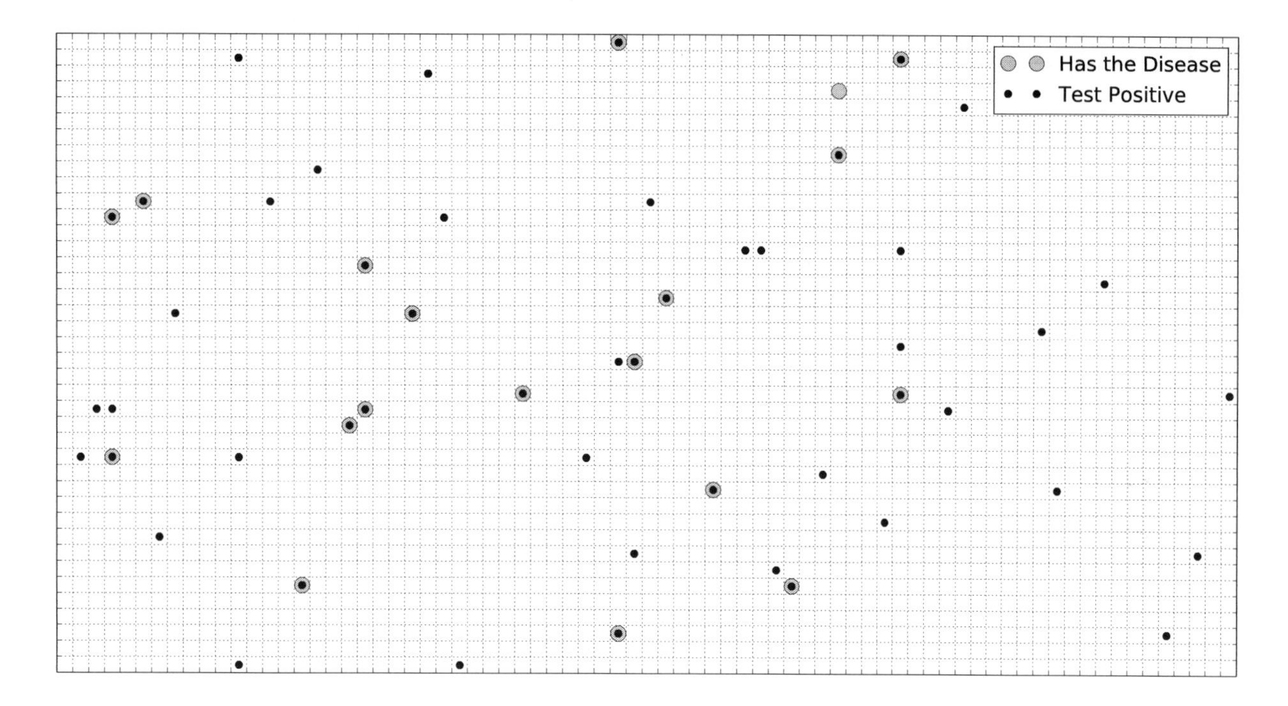

Figure 5.1: Rare disease and testing. Shown is a population of 3000 where 1 in every 200 people have the disease (large circles). A test which is 99% effective is applied to everyone in the population, and the positive test results (i.e. the test says that you have the disease) are shown ask small black dots. Notice that although nearly all of those that have the disease test positive (a small black dot inside a large circle), there are many false positives (black dot in an empty square) - healthy people that test positive for the disease. Even though the test is quite good, there are many more healthy people and 1 out of 100 of them will erroneously test positive.

Consequences

This sort of disease testing has serious consequences, especially for rare diseases with test that aren't precise. In the book "The Theory That Would Not Die: How Bayes' Rule Cracked the Enigma Code, Hunted Down Russian Submarines, and Emerged Triumphant from Two Centuries of Controversy" by Sharon McGrayne there is a discussion concerning the 2009 advice from the U.S. government task force that "most women in their forties *not* to have annual mammograms." (emphasis mine) According to McGrayne,

> Thus the probability that a woman who tests positive has breast cancer is only 3%. She has 97 chances out of 100 to be disease free. None of this is static. Each time more research data become available, Bayes' rule should be recalculated. As far as Bayes is concerned, universal screening for a disease that affects only 4/10 of 1% of the population may *subject many healthy women to needless worry and to additional treatment which in turn can cause its own medical problems.* In addition, the money spent on universal screening could potentially be used for other worthwhile projects. Thus Bayes highlights the importance of improv-

ing breast cancer screening techniques and reducing the number of false positives.[1] (emphasis mine)

Thus the proper application of probability theory allows us to separate true but unintuitive things from others which only seem true and intuitive but are in fact false.

[1] Sharon McGrayne. *The Theory That Would Not Die: How Bayes' Rule Cracked the Enigma Code, Hunted Down Russian Submarines, and Emerged Triumphant from Two Centuries of Controversy*. Yale University Press, 2011. ISBN 0300169698

5.2 *Monty Hall Problem*

This problem was introduced in Section 2.3.

EXAMPLE 5.1 *Is it better to switch doors? - Monty Hall Problem revisited*

You may recall that we were presented with a choice of 3 doors where a car is behind one and goats behind the others. Having picked one, the host opens up a door with a goat, and offers you the opportunity to change your answer. In order to assess the probabilities, we must remember that

1 the host *never* opens your door

2 the host *always* opens a door with a goat

We'll go through a specific example, that of you choosing door 1 and the host opening door 2. The analysis proceeds in identical ways for the other possibilities. We apply the Bayes' Recipe, where the models under consideration are

- "car behind door 1"
- "car behind door 2"
- "car behind door 3"

The Bayes' Recipe proceeds as follows

1 Specify the prior probabilities for the models being considered

$$\begin{aligned} P(\text{car } 1|\text{you } 1) &= 0.333 \\ P(\text{car } 2|\text{you } 1) &= 0.333 \\ P(\text{car } 3|\text{you } 1) &= 0.333 \end{aligned}$$

where, for example, $P(\text{car } 1|\text{you } 1)$ represents the probability that the door contains the car given that you chose door 1. Since your choice of door doesn't add any information about the location of the car, all of the probabilities are equal.

2 Write the top of Bayes' Rule for all models being considered

$$\begin{aligned} P(\text{car } 1|\text{you } 1, \text{host } 2) &\sim P(\text{host } 2|\text{you } 1, \text{car } 1)\, P(\text{car } 1|\text{you } 1) \\ P(\text{car } 2|\text{you } 1, \text{host } 2) &\sim P(\text{host } 2|\text{you } 1, \text{car } 2)\, P(\text{car } 2|\text{you } 1) \\ P(\text{car } 3|\text{you } 1, \text{host } 2) &\sim P(\text{host } 2|\text{you } 1, \text{car } 3)\, P(\text{car } 3|\text{you } 1) \end{aligned}$$

3 Put in the likelihood and prior values

Due the restrictions on the host above, the host cannot open a door with a car, so $P(\text{host } 2|\text{you } 1, \text{car } 2) = 0$. In the case where you choose door 1 and the car is also behind door, the host has the freedom to choose either door 2 or door 3, so $P(\text{host } 2|\text{you } 1, \text{car } 1) = 0.5$. Where the information comes in is when the car is behind door 3 and you've chosen door 1. In that case, the host cannot open your door (door 1) or the door with the car (door 3) and *must* open door 2. Thus, $P(\text{host } 2|\text{you } 1, \text{car } 3) = 1$.

The final result of this step is

$$\begin{aligned} P(\text{car } 1|\text{you } 1, \text{host } 2) &\sim 0.5 \cdot 0.333 \\ P(\text{car } 2|\text{you } 1, \text{host } 2) &\sim 0 \cdot 0.333 \\ P(\text{car } 3|\text{you } 1, \text{host } 2) &\sim 1 \cdot 0.333 \end{aligned}$$

4 Add these values for all models

$$K = 0.5 \cdot 0.333 + 1 \cdot 0.333 = 0.5$$

5 Divide each of the values by this sum, K, to get the final probabilities

$$\begin{aligned} P(\text{car } 1|\text{you } 1, \text{host } 2) &= \frac{0.5 \cdot 0.333}{0.5} = 0.333 \\ P(\text{car } 2|\text{you } 1, \text{host } 2) &= \frac{0 \cdot 0.333}{0.5} = 0 \\ P(\text{car } 3|\text{you } 1, \text{host } 2) &= \frac{1 \cdot 0.333}{0.5} = 0.666 \end{aligned}$$

Thus, in the case, given that you choose door 1 and the host chooses 2, the probability that the car is behind door 1 (your door) is 0.333 and the other door (door 3) is 0.666. Following the same steps through the other cases, we get in summary

		Probability of...		
Your Choice	Host Choice	Car Behind 1	Car Behind 2	Car Behind 3
1	1	(host can't open your door)		
1	2	0.333	0	0.666
1	3	0.333	0.666	0
2	1	0	0.333	0.666
2	2	(host can't open your door)		
2	3	0.666	0.333	0
3	1	0	0.666	0.333
3	2	0.666	0	0.333
3	3	(host can't open your door)		

In summary, it is *always* better to switch to the remaining door, given these rules.

5.3 *Psychic Octopi*

There was a German octopus named Paul[2] who was claimed to be psychic during his lifetime. He was given this designation because he was supposedly able to pick the result of World Cup matches before they occurred[3]. His impressive results, across 2 years, shown in Figure 5.2 can be summarized as follows:

$$\text{data} \equiv \text{12 out of 14 correctly predicted}$$

[2] Paul the octopus, July 2012. URL http://en.wikipedia.org/wiki/Psychic_octopus

[3] The basic procedure for Paul to make a "prediction" was for his trainers to present two food dishes, labeled with a flag representing the two countries, respectively, competing. Whichever food dish Paul chose first was his prediction for the winner of the game.

UEFA Euro 2008

Teams	Stage	Date	Prediction	Result	Outcome
Germany vs Poland	Group stage	8 June	Germany	2–0	Correct
Croatia vs Germany	Group stage	12 June	Germany	2–1	Incorrect
Austria vs Germany	Group stage	16 June	Germany	0–1	Correct
Portugal vs Germany	Quarter-finals	19 June	Germany	2–3	Correct
Germany vs Turkey	Semi-finals	25 June	Germany	3–2	Correct
Germany vs Spain	Final	29 June	Germany[6]	0–1	Incorrect

2010 FIFA World Cup

Teams	Stage	Date	Prediction	Result	Outcome
Germany vs Australia	Group stage	13 June	Germany[22]	4–0	Correct
Germany vs Serbia	Group stage	18 June	Serbia[22]	0–1	Correct
Ghana vs Germany	Group stage	23 June	Germany[22]	0–1	Correct
Germany vs England	Round of 16	27 June	Germany[23]	4–1	Correct
Argentina vs Germany	Quarter-finals	3 July	Germany[24]	0–4	Correct
Germany vs Spain	Semi-finals	7 July	Spain[25]	0–1	Correct
Uruguay vs Germany	3rd place play-off	10 July	Germany[26]	2–3	Correct
Netherlands vs Spain	Final	11 July	Spain[27]	0–1	Correct

Figure 5.2: The full results of the predictions of Paul the Octopus, reproduced from en.wikipedia.org/wiki/Psychic_octopus.

The question we have to ask is, is this data strong evidence for a psychic octopus? In order to have a well-posed problem we need the following three components:

1. a set of hypotheses, or models, to compare - we need at least two, otherwise the question is meaningless

2. for each model, an equation denoting the *likelihood*, or in other words, how probable is the data given the particular model

3. a specification of the *prior* probability, or in other words, how likely was our model before we saw the data

Making a Well Posed Problem

We are interested in the probability of this octopus being psychic, given this data, or

$$P(\text{psychic}|\text{data})$$

which really is an example of a model comparison, or hypothesis testing. In any kind of model comparison, we need to have *multiple models* to compare to in order to proceed. The models we consider constrain the problem, and define which ideas we are willing to consider. To be specific, as a first step, let's consider the following two models

$$\begin{aligned} H &:= \{\text{Paul is psychic}\} \\ R &:= \{\text{Paul is completely random, like a coin flip}\} \end{aligned}$$

The next step is to be able to assign probabilities from these models. It is easy for the *random* hypothesis

$$\begin{aligned} P(\text{correct prediction}|R) &= 0.5 \\ P(\text{incorrect prediction}|R) &= 0.5 \end{aligned}$$

What does it mean to be psychic? What is the probability of getting a correct result if you are psychic? According to James Randi[4] many of the psychics and dowsers claim 100% accuracy in their predictions before they are tested. However this would mean a *single* wrong answer would drive the probability of that model to *zero*: a perfect predictor cannot, logically, make any mistakes. For our case here, we choose to be generous to the psychic and allow for a reasonable failure rate, using 90% as our accuracy, thus

[4] J. Randi. *Flim-flam!: psychics, ESP, unicorns, and other delusions*, volume 342. Prometheus Books Amherst, NY, 1982

$$\begin{aligned} P(\text{correct prediction}|H) &= 0.9 \\ P(\text{incorrect prediction}|H) &= 0.1 \end{aligned}$$

Specifying the prior probability of these two models is a bit more challenging. It seems reasonable to assign a small prior probability to a psychic octopus - how many psychic octopi have you ever encountered? A small, but still quite conservative value, would be 1/100, so we have for the two models:

It is possible that we could be accused of an anti-psychic bias here, especially from someone who is a true believer. Why shouldn't the prior be $P(H) = 1/2$? If you had no world experience, that is what you'd start with, but then the behavior of the first octopi that you encounter would generally lower your assignment of the probability of the next octopi being psychic. After enough world experience, updating your probability with Bayes' Rule, you'd arrive at a very small prior for Paul, the current octopus we are examining.

$$\begin{aligned} P(H) &= 1/100 \\ P(R) &= 99/100 \end{aligned}$$

The First Model Comparison

Now that we've set up the problem, we can apply the Bayes' Recipe

1 Specify the prior probabilities for the models being considered

$$\begin{aligned} P(H) &= 1/100 \\ P(R) &= 99/100 \end{aligned}$$

2 Write the top of Bayes' Rule for all models being considered

$$\begin{aligned} P(H|\text{data} = \text{12 out of 14}) &\sim P(\text{data} = \text{12 out of 14}|H)P(H) \\ P(R|\text{data} = \text{12 out of 14}) &\sim P(\text{data} = \text{12 out of 14}|R)P(R) \end{aligned}$$

where we are using the symbol $\sim$ to denote *proportionality* or *related to*. Essentially, by calculating the top of Bayes' Rule first, the numbers are not *equal* to the final (i.e. posterior) probabilities but must be rescaled to make sure that they add up to 1. This is done in the final step. Up until that rescaling, we use the symbol $\sim$ and think of it as *related to*.

3 Put in the likelihood and prior values

$$\begin{aligned} P(H|\text{data} = \text{12 out of 14}) &\sim \binom{14}{12} 0.9^{12} 0.1^{14-12} \times \frac{1}{100} \\ &= 0.00257 \\ P(R|\text{data} = \text{12 out of 14}) &\sim \binom{14}{12} 0.5^{12} 0.5^{14-12} \times \frac{99}{100} \\ &= 0.00549 \end{aligned}$$

4 Add these values for all models

$$K = 0.00257 + 0.00549 = 0.00806$$

5 Divide each of the values by this sum, K, to get the final probabilities

$$\begin{aligned} P(H|\text{data}) &= \frac{0.00257}{0.00806} = 0.32 \\ P(R|\text{data}) &= \frac{0.00549}{0.00806} = 0.68 \end{aligned}$$

and the psychic loses!

Furthering the Comparison

Typically, a person who is supportive of psychic phenomena would choose a prior for our psychic hypothesis (H) that would be at least as large as the prior for the random hypothesis (R). In this case, the (posterior) probability of the octopus being psychic given the data of 12 correct out of 14 would be much higher. After "ruling out" the random octopus hypothesis, we'd be left with psychic. But is that all that is really left? No, and the analysis is easy to do.

Once presented with the success of Paul, most people instantly are suspicious of random octopus, but don't adopt psychic octopus as the answer. Perhaps the keepers, being German, biased the data taking a little bit. Perhaps the octopus chose flags with bright yellow stripes. Notice that each of these cases still results in the same data - the octopus would have gotten 12 out of 14, but the *prior* probability of these cases should be much higher than psychic, even if lower than random. We leave it as an exercise to perform the calculation in this case, but it is directly parallel to the *Nines* deck example of Section 4.2 on page 96.

6 *Introduction to Parameter Estimation*

We will introduce the idea of what is called *parameter estimation* using a simple system of bent coins. This will generalize to more complex models, and form the basis for much of statistical inference.

6.1 *Bent Coins*

Figure 6.1: Bent Coin

Imagine we have a series of coins bent by various amounts (Figure 6.1). If the coin is bent completely in half, then we could have the coin always flip heads (i.e. $P(\text{heads}) = 1$) or tails (i.e. $P(\text{tails}) = 1$) depending on how it is bent. If you don't bend the coin at all then we'd have a fair coin ($P(\text{heads}) = P(\text{tails}) = 0.5$). So, let's say that we have a collection of bent coins which are bent by different amounts. For convenience we will number them from 0 to 10. The Table 6.1 summarizes the probability of each coin flipping heads.

Why do we number them from zero here? It's so that the number of the coin, say number 7, corresponds the probability that that coin flips heads, $P(\text{heads}) = 0.7$

Table 6.1: Probabilities for flipping heads given a collection of bent coins

Coin Number	Probability for Flipping Heads ($P(\text{heads})$)
0	0.0
1	0.1
2	0.2
3	0.3
4	0.4
5	0.5
6	0.6
7	0.7
8	0.8
9	0.9
10	1.0

Now I have the following scenario[1], with a few questions.

Imagine I have taken a random coin from my collection, flipped it and got the following data:

T T T H T H T T T T T H (i.e. 9 tails and 3 heads)

1 From this data, which coin do I most likely have?

[1] D. V. Lindley and L. D. Phillips. Inference for a bernoulli process (a bayesian view). *The American Statistician*, 30(3):112–119, 1976

2 Can we be *significantly confident* that this particular coin will result in more tails than heads in the future?

The way we've set up this problem is exactly like the model comparison example with the High and Low Deck (Section 4.1), except in this case we have 11 models (one for each coin). Applying the Bayes' Recipe we have

1 Specify the prior probabilities for the models being considered. Given no further information, we select a *uniform* distribution for the prior (i.e. all models are initially equally probable):

$$\begin{aligned} P(M_0) &= 1/11 \\ P(M_1) &= 1/11 \\ &\vdots \\ P(M_{10}) &= 1/11 . \end{aligned}$$

where M_0 is the model defined by "we're flipping coin 0," M_1 is the model defined by "we're flipping coin 1," etc...

2 Write the top of Bayes' Rule for all models being considered:

$$\begin{aligned} P(M_0|\text{data}=9T,3H) &\sim P(\text{data}=9T,3H|M_0)P(M_0) \\ P(M_1|\text{data}=9T,3H) &\sim P(\text{data}=9T,3H|M_1)P(M_1) \\ &\vdots \\ P(M_{10}|\text{data}=9T,3H) &\sim P(\text{data}=9T,3H|M_{10})P(M_{10}) . \end{aligned}$$

3 Put in the likelihood and prior values. Here we are drawing from a *binomial* distribution for the likelihood:

$$\begin{aligned} P(M_0|\text{data}=9T,3H) &\sim \binom{12}{3} 0.0^3 \times (1-0.0)^9 \times 1/11 \\ P(M_1|\text{data}=9T,3H) &\sim \binom{12}{3} 0.1^3 \times (1-0.1)^9 \times 1/11 \\ &\vdots \\ P(M_{10}|\text{data}=9T,3H) &\sim \binom{12}{3} 1.0^3 \times (1-1.0)^9 \times 1/11 . \end{aligned}$$

4 Add these values for all models: see Table 6.2.

5 Divide each of the values by this sum, K, to get the final probabilities: see Table 6.2.

When we are dealing with this many models, it is easier to plot the results, shown in Figure 6.2. We are now in a position to address the questions posed at the beginning of the section.

Model	$\sim P(M_i \vert \text{data} = 9T, 3H)$	$\sim P(M_i \vert \text{data} = 9T, 3H)/K$
M_0	0.000	0.000
M_1	0.00774	0.110
M_2	0.0214	0.306
M_3	0.0217	0.310
M_4	0.0128	0.184
M_5	0.00488	0.0696
M_6	0.00113	0.0161
M_7	0.000135	0.00192
M_8	0.00000524	0.0000748
M_9	0.0000000145	0.000000208
M_{10}	0.000	0.000
	K=0.0700	

Table 6.2: Probability for different bent-coin models, given the data=9 tails, 3 heads. The middle column is the non-normalized value from Bayes' Rule, needing to be divided by K (the sum of the middle column) to get the final column which is the actual probability.

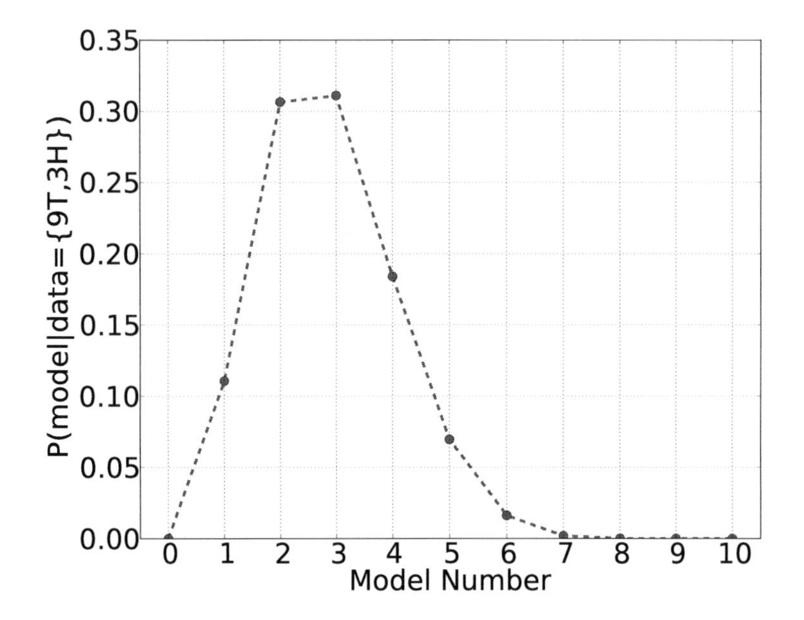

Figure 6.2: Probability for different bent-coin models, given the data=9 tails, 3 heads.

1 From this data, which coin do I most likely have?

   The maximum probability is for coin 3, but coin 2 is a close second. Thus we can be reasonably confident that we have been flipping one of those two coins, but can't narrow our confidence any more than that.

2 Can we be *significantly confident* that this particular coin will result in more tails than heads in the future?

   This is another way of asking for the total probability for coins less than coin 5 (the fair coin), or

$$P(\text{coin } 0 \textbf{ or } \text{coin } 1 \textbf{ or } \text{coin } 2 \textbf{ or } \text{coin } 3 \textbf{ or } \text{coin } 4) = \\ 0.000 + 0.110 + 0.306 + 0.310 + 0.184 = 0.912$$

   which says that this coin is "likely" to "very likely" (Table 1.1 on page 45) to have a probability of yielding heads less than a fair coin, and thus yield more tails in the future.

6.2 *Priors versus Data*

It is instructive to pause and look at this example one flip at a time, to see how the probability and thus our state of knowledge adjusts as we collect more data. In Figure 6.3 we see the result of our procedure when there is no data (i.e. our initial, prior probabilities) and when we've flipped once and then again, both times tails. The curve for no data is the same as the prior probability, and in this case all models are equally likely. When the first tails is observed, the model which states that heads are *certain* (i.e. coin 10) goes to zero probability because coin 10 *cannot* flip tails.[2]. At this point we *know* that it is *impossible* for us to be flipping coin 10. We see also that the high-numbered coins (i.e. the ones with high probability of flipping heads) have greatly reduced probability while we've seen only tails.

As more tails are observed, the probability for the lower models is increased. As we flip more tails we become more confident in the lower-number models. Because at this point we haven't flipped any heads, the model 0 still has non-zero probability - it is still possible that we are holding a coin that cannot flip heads.

When we continue with the next few flips (Figure 6.4) we encounter our first heads on the fourth flip. At this point the model which states that heads are *impossible* (i.e coin 0) goes to *zero* probability. Finally, across our entire data set (Figure 6.5) we see that the curve gets narrower, where more of the probability falls on only a few of the models and the other models become less and less likely. With only 12 data points, there is still a lot of uncertainty in which

[2] Notice that the only models with probability equal to zero are ones that are *logically impossible*. It's not the colloquial usage of impossible, as in "it is impossible for the Red Sox to win this year," but in the strict usage, as in "it is impossible to flip both heads and tails at the same time." The reason this is the case is that a statement with zero probability cannot be made possible with *any about of data* - it is an utterly *dogmatic* statement. Thus, we reserve it only for things that are *logically impossible*.

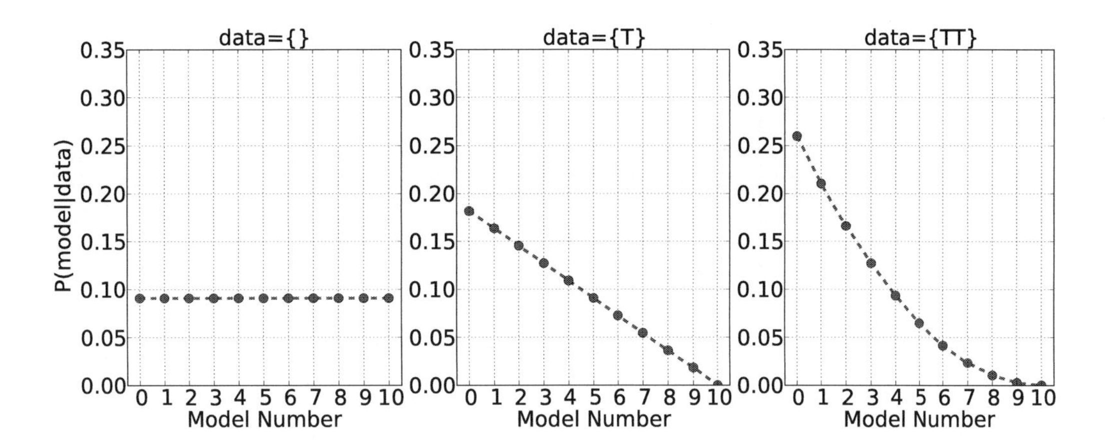

Figure 6.3: Probability for different bent-coin models, given no data (left), the first tails (middle), and the second tails (right). The curve for no data is the same as the prior probability, and in this case all models are equally likely. When the first tails is observed, the model which states that heads are *certain* (coin 10) goes to zero probability. As more tails are observed, the probability for the lower models is increased.

model - several models have reasonably high probability values. We still can rule out a few models confidently (like coins 0, 6, 7, 8, 9, and 10). We are most confident in coins 2 and 3, with the most probability.

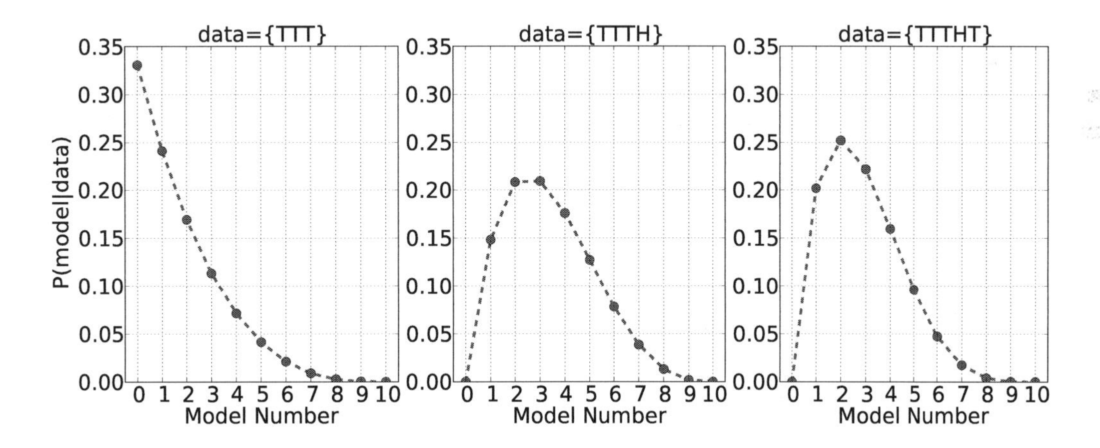

Figure 6.4: Probability for different bent-coin models, given three tails (left), the first heads (middle), and another tails (right). When the first heads is observed, the model which states that heads are *impossible* (coin 0) goes to zero probability.

6.3 *Moving Toward the Continuous*

There is a practical problem that we face at this point, when we consider a generic bent coin. Perhaps it doesn't fit in one of the 11 models considered, falling somewhere in between, for example with

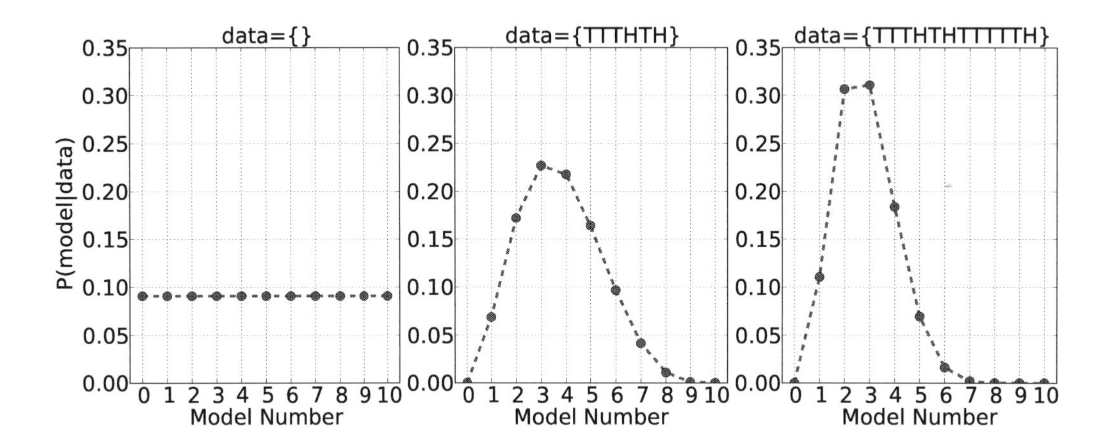

Figure 6.5: Probability for different bent-coin models, given no data (left), the first half of the data set (middle), and the entire data set of 9 tails and 3 heads (right).

$P(\text{heads}) = 0.132464$. Ones' first thought might be to include one thousand coins or one million coins instead of the 11 we've considered so far, so we could have coin 132464, coin 132465, coin 132466, etc... Although this can be done, we run into two problems

1 Because we are dealing with so many models, the probability associated with any *single* model gets very small - and gets smaller with the more models you consider

2 We can't practically distinguish between models such as $P(\text{heads}) = 0.132464$ and $P(\text{heads}) = 0.13246\mathbf{5}$ (the last digit is different here)

In order to solve both of these problems mathematically, we introduce the concept of a *continuous distribution*. We start by labeling the model with a *continuous* number rather than an integer. In our present case it makes sense to label the model with the probability that the coin flips heads. We'll call this label θ, and it will have a value between 0 (heads are impossible) and 1 (heads are certain) and can take on *any value* in between. Because we now have an infinite number of labels, we have two consequences:

1 We can't simply add up all the probabilities to get our value of K to make everything add up to 1. Instead, we look at *areas under the curve* and make sure the entire area equals 1.

2 Because, with distributions, *areas under the curve* (and not the values of the distribution itself) are the probabilities, we can only speak about *ranges* of values. For example, we can speak meaningfully about the probability of θ between 0.3 and 0.4 (i.e. $P(0.3 < \theta < 0.4)$). When we write down something like $P(\theta) = 1$ we're not talking about a probability of a single label but rather the magnitude of the distribution at that label, θ.

We revisit Bayes' Recipe again, using the distributions. This time we also will look at pictures of the distributions as we progress.

1 Specify the prior probabilities for the models being considered:

$$P(\theta) = 1\,.$$

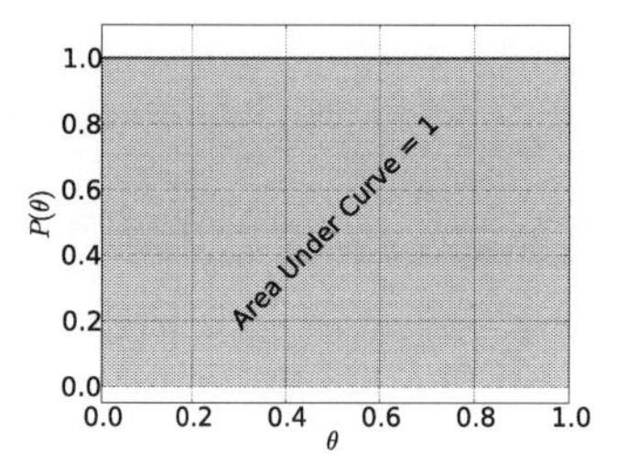

2 Write the top of Bayes' Rule for all models being considered:

We can write one equation for all of the models labeled by θ at once as

$$P(\theta|\text{data} = 9T,3H) \quad \sim \quad P(\text{data} = 9T,3H|\theta)P(\theta)\,.$$

3 Put in the likelihood and prior values.

We use the binomial model, one equation for all models, remembering that for a model labeled by θ the probability for that coin flipping heads is $P\,(\text{heads}) = \theta$. Thus we get the likelihood and prior values as

$$\begin{aligned} P(\theta|\text{data}) \quad &\sim \quad P(\text{data}|\theta) \cdot P(\theta) \\ P(\theta|\text{data}9T,3H) \quad &\sim \quad \binom{12}{3} \theta^3 \times (1-\theta)^9 \cdot 1 \end{aligned}$$

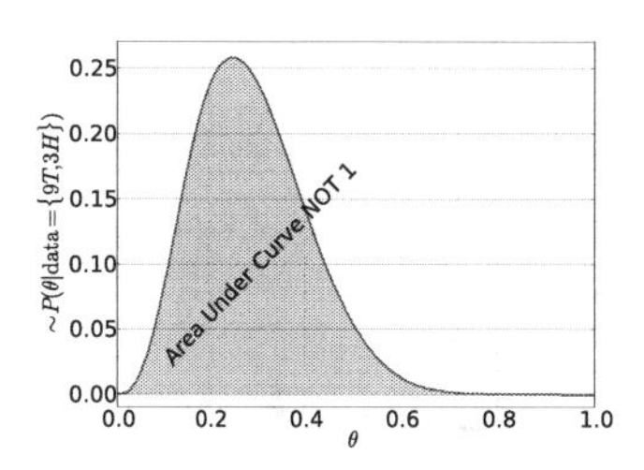

4 Find the area under this curve, and call it K.

5 Divide each of the values of the curve by this are, K, to get the final probabilities where the area under the curve is 1.

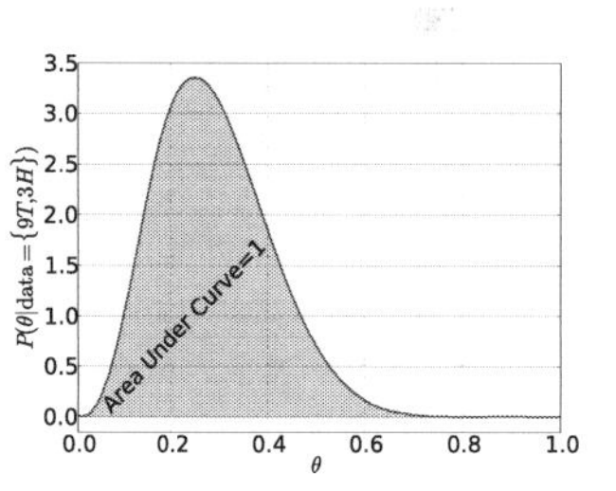

Usually these steps are done for you, for a specific data set, and you are given the final posterior distribution to use in answering any questions. However, for any particular case it is important to know what assumptions have been made in the choice of models and model parameters.

6.4 *MAP and Areas*

Now we revisit the questions posed in Section 6.1 on page 113 about the bent coin, this time using the distribution found above, reproduced here in Figure 6.6.

> Imagine I have taken a random coin from my collection, flipped it and got the following data:
>
> T T T H T H T T T T T H (i.e. 9 tails and 3 heads)

1 From this data, which "coin" do I most likely have? (or in this interpretation, what is my best estimate for the probability of this coin flipping heads, denoted by θ)

2 Can we be *significantly confident* that this particular coin will result in more tails than heads in the future?

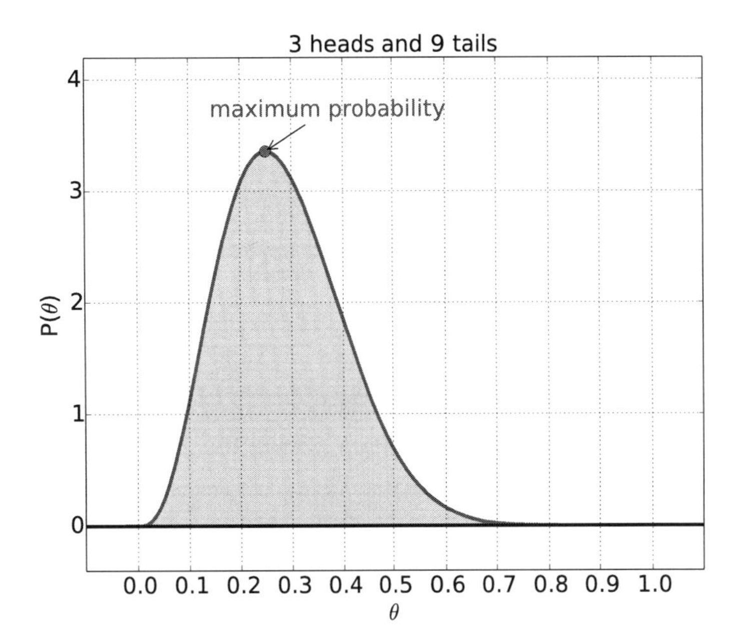

Figure 6.6: Posterior probability distribution for the θ values of the bent coin - the probability that the coin will land heads. The distribution is shown for data 3 heads and 9 tails, with a maximum at $\theta = 0.25$.

One answer to the first question can be accomplished by looking at the *maximum* of the posterior distribution, shown in Figure 6.6.[3] By eye, it seems to have a maximum 0.25. In fact one can demonstrate that this distribution has a maximum at

$$\theta_{\text{max}} = \frac{\text{number of successes}}{\text{total number of attempts}},$$

where in our example, a success is head, and an attempt is a flip.[4] We take up this question of the best estimate of θ, given the posterior probability for θ, in more detail in Section 6.6.

The answer to the second question can be done by looking at the area under the curve from $\theta = 0$, the "all heads" coin, to $\theta = 0.5$, the "fair" coin, as shown in Figure 6.7. This area represents the probability, given the data, that the coin is skewed towards heads or, in other words, how confident are we that this is an unfair coin. Given the value of $P(\theta < 0.5) = 0.954$ we can say that this is "very likely" an unfair coin (see Table 1.1 on page 45).

[3] The maximum of the posterior distribution, which represents the most likely value of a quantity, is often referred to as the *MAP estimate*. It is also commonly referred to as the *mode* of the distribution.

[4] This distribution, given how common it is, is given the name *Beta distribution*. There are a handful of common distributions that are given names for convenience. We've already seen the *uniform distribution*, and there will be others.

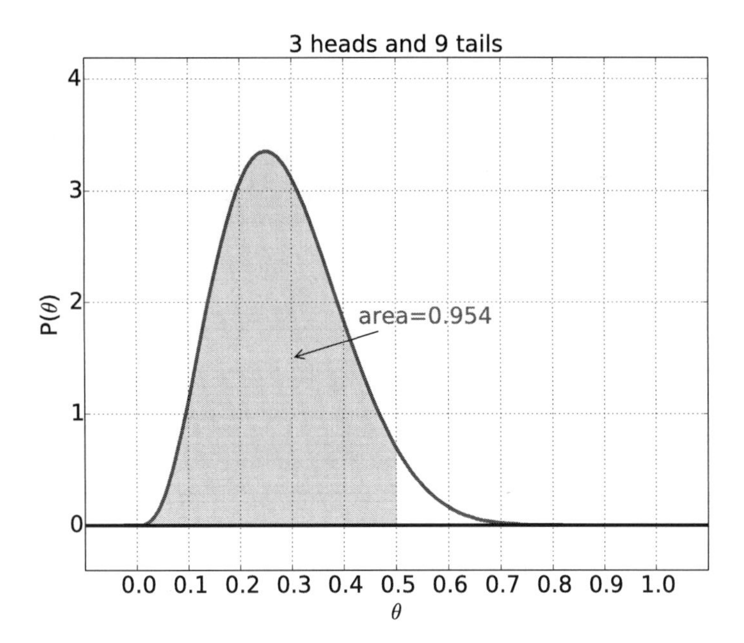

Figure 6.7: Posterior probability distribution for the θ values of the bent coin - the probability that the coin will land heads. The distribution is shown for data 3 heads and 9 tails. The area under the curve from $\theta = 0$ (the "all heads" coin) to $\theta = 0.5$ (the "fair" coin) is 0.954.

6.5 *Quartiles*

Given that we are dealing most often with continuous distributions, and thus need to look at areas under the curve from one point to another, it is useful to make a table for a distribution of these areas. Typically we look at the values of the parameter at which we have a given area under the curve from the minimum possible value of the parameter up to to that value. For example, we might be interested in the value of θ (i.e. how skewed the coin is) such that we have an area of 50% from 0 up to θ, shown in Figure 6.8. This point (called the *median*) represents the point where we would be just as confident (given our data) that the coin is *more* skewed than this as *less* skewed.

A table of these values for a distribution can be very useful. For example, consider the table and plot shown in Figure 6.9. Shown are the various points where the area under the curve up to those points is specified. For example, the area under the curve from $\theta = 0$ up to $\theta = 0.11$ is 5%. This means, given the data of 3 heads and 9 tails, there is a probability $P = 5\%$ of the coin having less than $\theta = 0.11$, or an extreme skew towards tails.

Quartiles The term *quartiles* refers to the values of the parameter which result in an area of 25%, 50%, or 75%, or one, two, or three quarters of the area.

Quartiles The term *quartiles* refers to the values of the parameter which result in an area of 25%, 50%, or 75%, or one, two, or three quarters of the area.

When we wish to refer to a non-quarter percentage, then we'll call

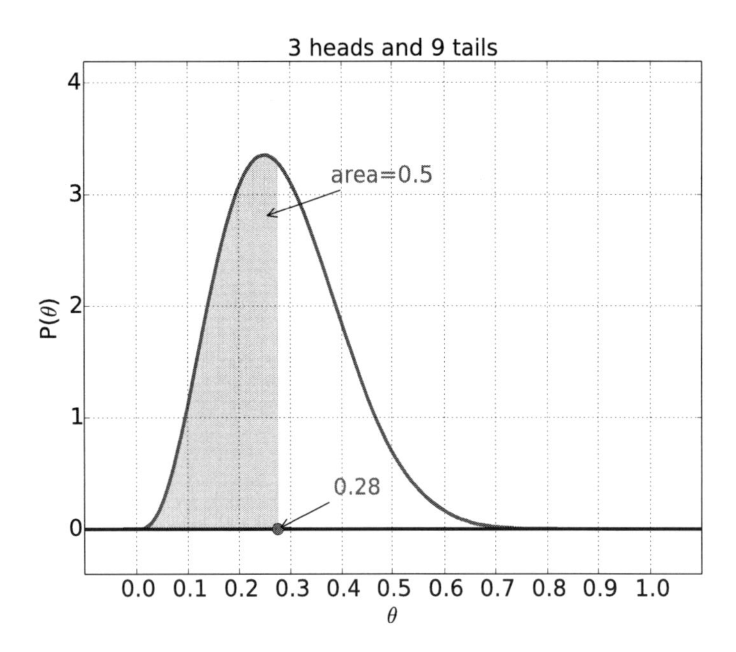

Figure 6.8: Posterior probability distribution for the θ values of the bent coin - the probability that the coin will land heads. The distribution is shown for data 3 heads and 9 tails. The area under the curve from $\theta = 0$ (the "all heads" coin) to $\theta = 0.28$ is 0.5 - half the area. This represents the *median* of the distribution.

it a *percentile*.

Percentiles The term *percentile* refers to the value of the parameter which result in a particulare area under the curve.

Percentiles The term *percentile* refers to the value of the parameter which result in a particulare area under the curve.

For example, we can say from Figure 6.9 that the 99% percentile is 0.59. Thus, it is extremely unlikely to have the coin skewed towards heads more than $\theta = 0.59$ given the observation that we flipped 3 heads and 9 tails with this coin.

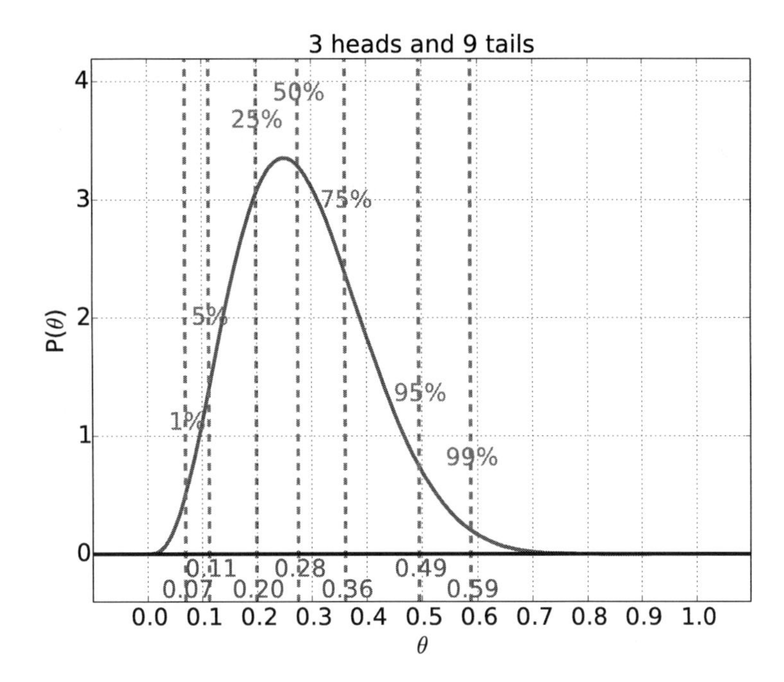

Beta(heads=3,tails=9)	
Value	**Area**
0.07	0.01
0.11	0.05
0.14	0.10
0.20	0.25
0.28	0.50
0.36	0.75
0.44	0.90
0.49	0.95
0.59	0.99

Figure 6.9: Posterior probability distribution for the θ values of the bent coin - the probability that the coin will land heads. The distribution is shown for data 3 heads and 9 tails. The various quartiles are shown in the plot, and summarized in the accompanying table.

6.6 *Best Estimates*

Perhaps surprisingly, there is not a single answer to the best estimate for θ given the poster distribution, for example the one shown in Figure 6.9. There are several plausible measures, each with their own advantages. Any specific estimate of a parameter (e.g. θ) is denoted with a hat (e.g. $\hat{\theta}$) in the descriptions that follow.

The Mode Also known as the *maximum a-posteriori probability* (MAP) estimate, the *mode* is the maximum of the posterior probability. In the case of a Beta distribution with h successes in N trials, we have

$$\hat{\theta}_{\text{mode}} = \frac{h}{N}$$

The Mode Also known as the *maximum a-posteriori probability* (MAP) estimate, the *mode* is the maximum of the posterior probability.

The Mean Also known as the *expected value* or *average value*, the *mean* of a distribution of a parameter θ is defined to be the sum of all of the possible values of θ times the posterior probability of θ,

$$\hat{\theta}_{\text{mean}} = \sum_{\theta} \theta \times P(\theta|\text{data})$$

It is one measure of the *middle* of the distribution. In the special case of a Beta distribution with h successes in N trials, we have

$$\hat{\theta}_{\text{mean}} = \frac{h+1}{N+2}$$

The Mean Also known as the *expected value*, the *mean* of a distribution of a parameter θ is defined to be the sum of all of the possible values of θ times the posterior probability of θ, as in

$$\hat{\theta}_{\text{mean}} = \sum_{\theta} \theta \times P(\theta|\text{data})$$

It is one measure of the *middle* of the distribution.

Intuitively this is the same as the MAP of the Beta distribution, with one more success and one more failure than actually observed. Further, for the Beta distribution, the mean value $\hat{\theta}_{\text{mean}}$ represents the predictive probability of a successful event on the *next* observation.

The Median Also known as the 50%-percentile, the median represents the middle of the distribution such that the probability of the parameter below the median equal to the probability of the parameter above the median.

$$P(\theta \leq \hat{\theta}_{\text{median}}|\text{data}) = P(\theta \geq \hat{\theta}_{\text{median}}|\text{data}) = 0.5$$

"Assume 2 successes and 2 failures" median approximation For the Beta distribution there is no simple form for the median, but a decent approximation which we will use is given by[5]

$$\hat{\theta}_{\text{median}} \approx \frac{h+2}{N+4}$$

Intuitively this is the same as the MAP of the Beta distribution, with two more successes and two more failures than actually observed, and is thus referred to as the "Assume 2 successes and 2 failures" median approximation.

Although each of these has their advantages, most notably ease of computation (especially for the mode and the mean), we will typically use the median of the distribution as the best estimate for the following reasons:

1 the median is intuitive as literally the middle of the distribution

2 the median is not as sensitive to distributions that are highly asymmetric

In most practical examples it may not make much difference, and for some distributions (such as the Normal distribution described in Chapter 7 (*Priors, Likelihoods, and Posteriors*)) there is not difference - the mean *is* the median which is also the mode.

The Median Also known as the 50%-percentile, the median represents the middle of the distribution such that the probability of the parameter below the median equal to the probability of the parameter above the median.

$$P(\theta \leq \hat{\theta}_{\text{median}}|\text{data}) = 0.5$$
$$P(\theta \geq \hat{\theta}_{\text{median}}|\text{data}) = 0.5$$

"Assume 2 successes and 2 failures" median approximation For the Beta distribution there is no simple form for the median, but a decent approximation which we will use is given by

$$\hat{\theta}_{\text{median}} \approx \frac{h+2}{N+4}$$

Intuitively this is the same as the MAP of the Beta distribution, with two more successes and two more failures than actually observed, and is thus referred to as the "Assume 2 successes and 2 failures" median approximation.

[5] Alan Agresti and Brian Caffo. Simple and effective confidence intervals for proportions and differences of proportions result from adding two successes and two failures. *The American Statistician*, 54 (4):280–288, 2000

EXAMPLE 6.1 *What is the best estimate of the probability of a bent coin flipping heads, given the observation of 9 tails and 3 heads?*

If we take the best estimate to be the median, then we have from the "assuming 2 successes and 2 failures" method,

$$\begin{aligned}\hat{\theta}_{\text{median}} &\approx \frac{h+2}{N+4} \\ &= \frac{5}{16} = 0.313\end{aligned}$$

Notice that the maximum probability was at the somewhat lower value

$$\hat{\theta}_{\text{mode}} = \frac{h}{N} = \frac{3}{12} = 0.25$$

One reason why the median is a better estimate in this case is because, as shown in Figure 6.9, there is more probability (i.e. area under the curve) to the right of the maximum than to the left, so the best estimate should be greater than the one given by the mode.

6.7 *Uncertainty in the Best Estimates*

To quantify the uncertainty in the best estimates, we need a value which represents the *width* of the distribution. Looking at Figure 6.10 we'd like to provide a quick way of saying that the range of probable values lies somewhere between $\theta = 0.2$ and $\theta = 0.5$ - anything outside of this contributes only a small amount to the probability, or in other words, we are most confident that our best estimate of θ lies between those 0.2 and 0.5. Depending on the application, the symmetry of the distribution, and other practical factors one may see a few potential measures of the *width* of the distribution.

Inter-Quantile Range The Inter-Quantile Range (ICR) is the range between the 25% and 75% quartiles, and represents 50% of the probability.

In Figure 6.10, the Inter-Quantile Range range is [0.29,0.40].

Inter-Quantile Range The Inter-Quantile Range (ICR) is the range between the 25% and 75% quartiles, and represents 50% of the probability.

95% Credible Interval (CI) The 95% Credible Interval (CI) is the range between the 2.5% and 97.5% quantiles, and thus represents 95% of the probability. According to Table 1.1 on page 45, it is "very likely" that our best estimate lies in this range.

In Figure 6.10, the 95% Credible Interval is nearly [0.2,0.5].

95% Credible Interval (CI) The 95% Credible Interval (CI) is the range between the 2.5% and 97.5% quantiles, and thus represents 95% of the probability. According to Table 1.1 on page 45, it is "very likely" that our best estimate lies in this range.

Standard Deviation The standard deviation is a measure of the half-width of a distribution, most commonly used specifically with reference to the particular *Normal* distribution. This will be defined more precisely in Section 7.2 on page 132), and will thus not be defined in general here.

Standard Deviation The standard deviation is a measure of the half-width of a distribution, most commonly used specifically with reference to the particular *Normal* distribution. This will be defined more precisely in Section 7.2 on page 132), and will thus not be defined in general here.

An approximate value for the standard deviation for the Beta distribution is

$$\sigma \approx \sqrt{\hat{\theta}(1 - \hat{\theta})/N}$$

From Figure 6.10, and using the median as the best estimate, $\hat{\theta}$, we get

$$\sigma \approx \sqrt{0.34(1 - 0.34)/30} = 0.09$$

Standard Deviation to Uncertainty To convert this number to an uncertainty, it is a mathematical consequence that about 65% of the area is within 1 value of σ, 95% of the area is within 2 values of σ, and 99% of the area within 3 values.

Standard Deviation to Uncertainty To convert this number to an uncertainty, it is a mathematical consequence that about 65% of the area is within 1 value of σ, 95% of the area is within 2 values of σ, and 99% of the area within 3 values.

So, of for the approximate 95% CI for the case shown in Figure 6.10 is

$$[0.34 - 2 \cdot 0.09, 0.34 + 2 \cdot 0.09] = [0.16, 0.52]$$

a bit more conservative range (larger uncertainty) than is given by the direct method of quantiles, but it much easier to calculate.

6.8 *Marginalization*

In Section 1.4 we introduced the concept of marginalization, and in Section 2.7 we performed a discrete example of this. In that section it was seen as simply a consequence of the sum and product rules. It was a way of taking a probability that depended on several factors, and eliminating all but the single factor we're interested in. If we have a *continuous* distribution, this process involves calculus and we will not cover it in detail, but it is the same process. In the case of the distribution above, we have a distribution over a single variable, like $\text{Beta}(\theta|h,t)$. Imagine that we have a distribution that depends on *two* parameters,

$$\text{MyDist}(\theta, \xi)$$

which specifies the probability of an event given each combination of the parameters, θ and ξ. We'd have to do a three-dimensional plot to visualize this. Many times, however, we want just the probability of one of the single parameters. In those cases we will write

$$P(\theta) \sim [\text{MyDist}(\theta, \xi)]_{\text{marginalize over } \xi}$$

where we are "summing" over all the values of the other parameters, leaving the details to the mathematicians, and simply using the result.

Likewise we can *marginalize* the parameter θ to get the distribution of the other variable.

$$P(\xi) \sim [\text{MyDist}(\theta, \xi)]_{\text{marginalize over } \theta}$$

This becomes important in Chapter 7 and Chapter 8.

6.9 *Exercises*

EXERCISE 6.1 *Given the posterior shown in Figure 6.10 for 10 heads and 20 tails, answer the following:*

1 *The most likely estimate for the parameter θ. What does this mean?*

2 *Is it likely that this is a fair coin?*

3 *What is $P(0 \leq \theta \leq 0.3)$ approximately?*

4 *What is $P(0.2 \leq \theta \leq 0.35)$ approximately?*

5 *What is the* median *value? What are the quartiles?*

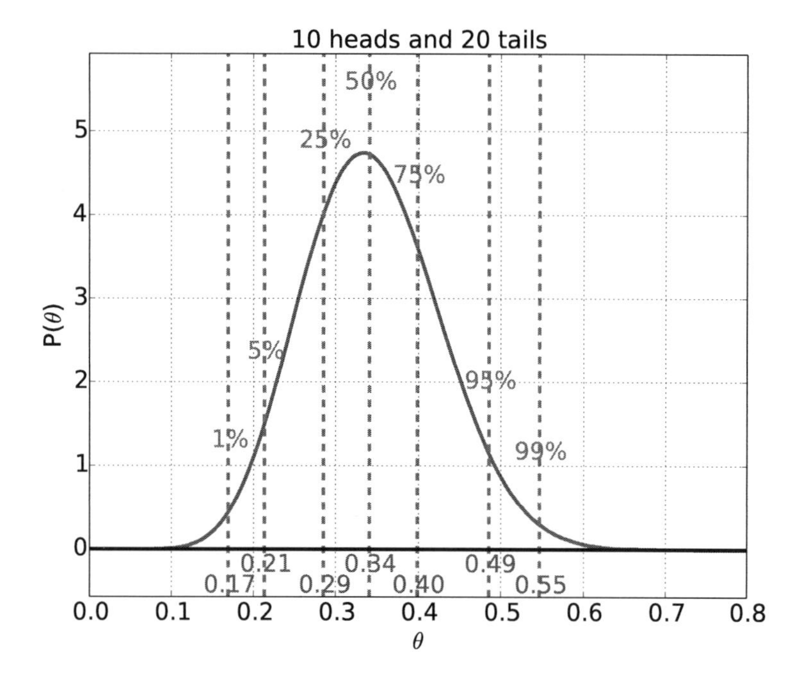

Beta(heads=10,tails=20)	
Value	**Area**
0.17	0.01
0.21	0.05
0.24	0.10
0.29	0.25
0.34	0.50
0.40	0.75
0.45	0.90
0.49	0.95
0.55	0.99

Figure 6.10: Posterior probability distribution for the θ values of the bent coin - the probability that the coin will land heads. The distribution is shown for data 10 heads and 20 tails. The various quartiles are shown in the plot, and

6.10 *Computer Examples*

```
from sie import *
```

Beta Distribution Example

3 heads and 9 tails Plot a beta distribution with 3 heads and 9 tails...

```
dist=beta(h=1,N=3)
distplot(dist,xlim=[0,1],show_quartiles=False)
```

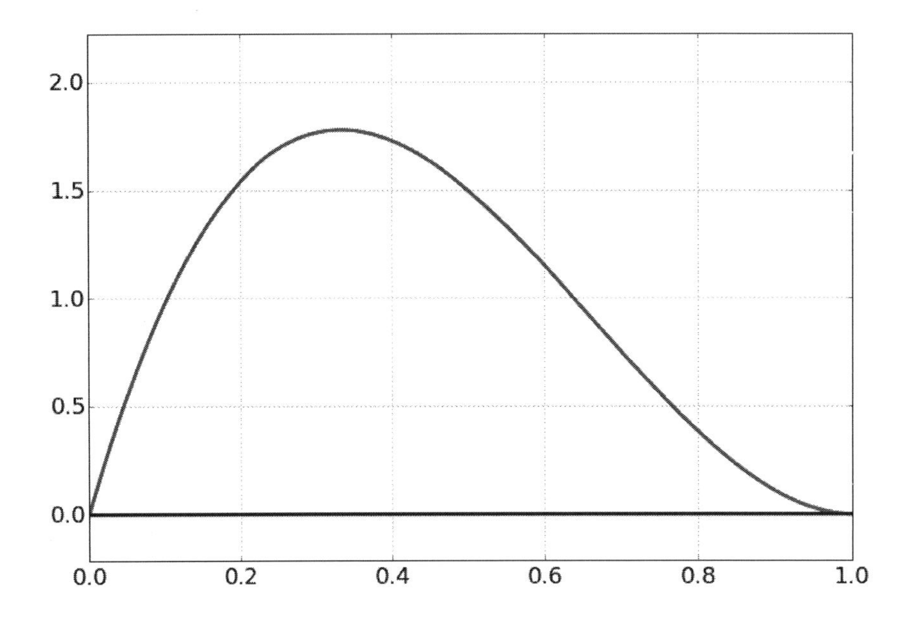

The median of this distribution...

```
dist.median()
```

```
0.27527583248615201
```

the 95

```
credible_interval(dist)
```

```
(0.067585986488542985, 0.38572756813238962, 0.80587955031675662)
```

1 heads and 3 tails This should be about the same fraction as the previous example, but broader

```
dist=beta(h=1,N=4)
distplot(dist,xlim=[0,1])
```

```
<matplotlib.figure.Figure at 0x108768cd0>
```

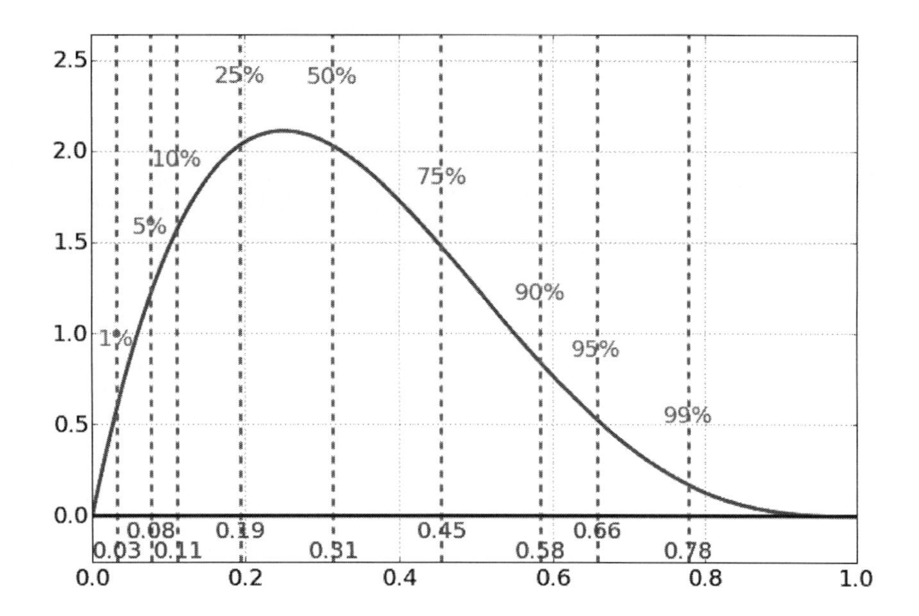


```
credible_interval(dist)
```

```
(0.052744950526316919, 0.31381017045569742, 0.71641793611808946)
```

```
```

7 Priors, Likelihoods, and Posteriors

7.1 Binomial and Beta Distributions

In Chapter 6 (*Introduction to Parameter Estimation* on page 113) we estimated the chance, θ, that a bent coin would come up heads by combining a *uniform prior* for θ (i.e. all possible values are a-priori equally likely) and a *binomial* likelihood (i.e. given θ, what is the probability of the data). This resulted in a *Beta* distribution for the posterior probability for θ.

Notice what the procedure of Bayes' Recipe is and how the Bayesian inference works here.

1 Specify the prior probabilities for the models being considered

> We want to estimate a quantity (which we label as θ), but begin with absolutely no knowledge of its value - we have a *uniform* prior probability.

2 Write the top of Bayes' Rule for all models being considered

> We construct a model for how different possible values of θ influence the outcome - a model we call the *likelihood*. In the case of the bent coin, the likelihood model is a *binomial* model, and describes the probability of flipping heads or tails given how bent the coin is (i.e. given θ).

3 Put in the likelihood and prior values

4 Add these values for all models

5 Divide each of the values by this sum, K, to get the final probabilities

> Once we observe data, we can combine the prior and the *model* or *likelihood* using the Bayes' recipe, and obtain the *posterior* distribution for the unknown value, θ, giving us the probability for each value, now updated with our new observations.

The last couple of steps of the recipe, for simple cases, is done by the mathematicians so we don't have to manually add and divide as we did in the previous chapters. In the case of the coin flips we get:

$$\underbrace{\text{Beta}(\theta|\text{data})}_{\text{posterior probability}} \sim \overbrace{\text{Binomial}(\text{data}|\theta)}^{\text{likelihood}} \times \underbrace{\text{Uniform}(\theta)}_{\text{prior probability}}$$

From this *Beta* distribution, we can get the most likely values (i.e. maximum probability value) for the unknown quantity of interest, θ, our *uncertainty* in this quantity (i.e. the *width* of the *Beta* distribution) consistent with the known data. In other words, the posterior probability summarizes all of our knowledge about the parameter of interest given the data.

7.2 *The Normal Distribution - Properties*

The *Normal* distribution, also referred to as the *Gaussian* distribution,[1] is by far the most commonly occurring distribution in all of statistical inference, so it requires some special attention.

[1] The distribution is named after Carl Friedrich Gauss who introduced it in 1809. However, it has been called in the past the Gauss-Laplacian distribution, due the the fact that Pierre Simone de Laplace was the first to apply it to real problems, and proved a number of very useful properties of it.

The Shape

The shape of the Normal distribution is sometimes described as *bell-shaped*, as shown in Figure 7.1, and is thus referred to as the bell-curve (although there are several other mathematical functions which are bell-shaped). The function is referred to as Normal(μ, σ) where μ and σ are parameters of the model. (see Appendix B.1 on page 183 for a review of greek letters)

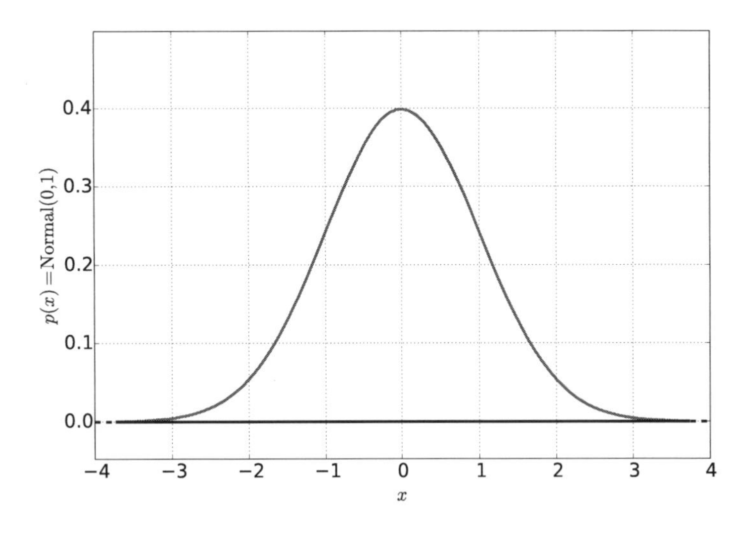

Figure 7.1: The Normal Distribution.

The location parameter, μ

The location parameter (see Figure 7.2) is the value of x for which the Normal distribution has a maximum probability. In a real sense, it is the *middle* of the distribution, and the best estimate of x. For the Normal distribution the location parameter, μ, is at once the mean, median and mode of the distribution.

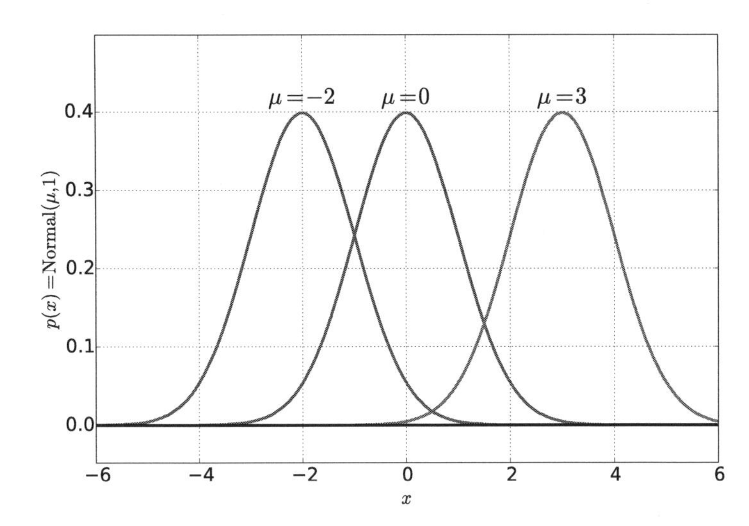

Figure 7.2: The Normal distribution with different location parameters, μ.

The deviation parameter, σ

As shown in Figure 7.3 the deviation parameter, σ, is a measure of how *spread out* the distribution is. As the width increases, the height goes down to keep the area under the curve constant (at 1). As a result, more of the probability sits at *larger* values of x as σ gets *larger*.

Three useful properties of σ for the Normal distribution are the following:

1. the Normal distribution value at the maximum (i.e. at $x = \mu$) is around 2.7 times larger than the value one-σ away from the maximum (at $x = \mu - \sigma$ and $x = \mu + \sigma$)

2. the total probability between these two points is 65%. This is typically written, $\mu \pm \sigma$.

3. 95% of the distribution lies between $\mu - 2\sigma$ and $\mu + 2\sigma$ (see Figure 7.3)

For example, writing 5 ± 2 typically implies a Normal distribution with mean $\mu = 5$ and deviation $\sigma = 2$. One is 65% certain that the range of the estimated value is between 3 and 7, and 95% certain that the range is between 1 and 9 (i.e. mean minus two deviations and mean plus two deviations).

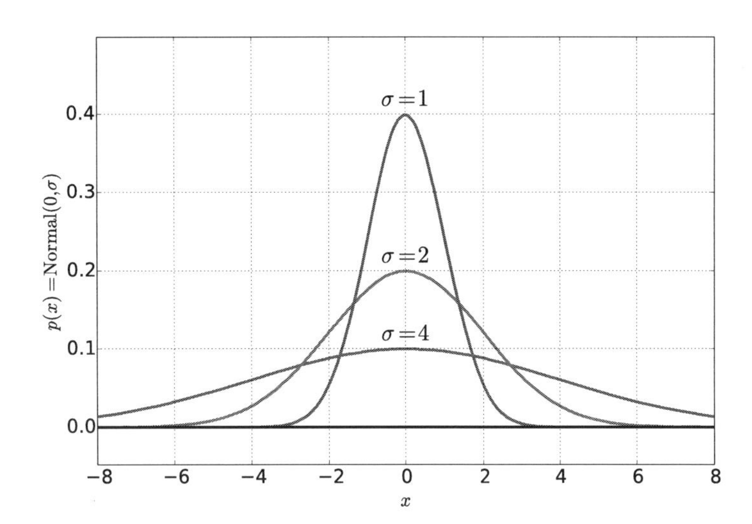

Figure 7.3: The Normal distribution with different deviation parameters, σ.

Summarizing the Distribution

We can specify the Normal distribution with just the two parameters, μ and σ - the location and deviation parameters, respectively. However, due to its symmetry, we can summarize this distribution for all cases by looking a a single special case called the *standard Normal distribution*.

The Standard Normal Distribution is the Normal distribution in the special case where $\mu = 0$ (the distribution is centered at $x = 0$) and $\sigma = 1$ (the distribution has a spread of 1).

The Standard Normal Distribution The Normal distribution in the special case where $\mu = 0$ (the distribution is centered at $x = 0$) and $\sigma = 1$ (the distribution has a spread of 1).

For any Normal distribution, the area within 1-σ is 0.68, within 2-σ is 0.95, and 3-σ is 0.99. These locations are the most prevalently used in any kind of statistical testing, and thus we will see them many times.

Moving from a General Normal to the Standard Normal and Back

In order to use the table of percentiles for the standard Normal distribution, we need to be able to translate from the Normal to the standard Normal and back again. Luckily, it is a simple process, and is one of the main reasons for using the Normal distribution - other distributions are not so easily manipulated.

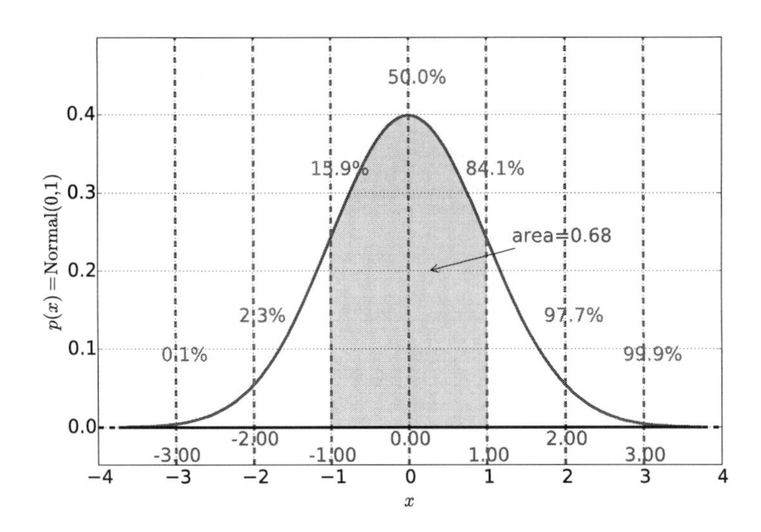

Figure 7.4: The Standard Normal Distribution (the Normal distribution in the special case where $\mu = 0$ and $\sigma = 1$). The percentiles shown are for positions 1-σ away from the center, 2-σ away, and 3-σ away. The area within 1-σ is 0.68, within 2-σ is 0.95, and 3-σ is 0.99. These locations are the most prevalently used in any kind of statistical testing, and thus we will see them many times.

To facilitate this translation, we will use the variable x for the Normal distribution and z for the standard Normal. So now, we need to have a recipe for translating x to z (or vice versa), given μ and σ. These recipes are:

1 x to z: subtract x by μ, and divide by σ

2 z to x: multiply z by σ and add μ

EXAMPLE 7.1 *Given a Normal distribution with a mean of* $\mu = 150$ *and a* $\sigma = 20$, *what is the most likely value?*

The most likely value is the peak of the probability distribution, $\hat{x} = \mu = 150$.

EXAMPLE 7.2 *Given a Normal distribution with a mean of* $\mu = 150$ *and* $\sigma = 30$, *what is the probability* $P(x > 170)$

To use the tables in Section D.3 on page 196, we first need to translate everything to the *standard* Normal values.

$$x = 170 \quad \Rightarrow \quad z = \frac{x - 150}{30} = 0.67$$

From the table in Section D.3 on page 196, the area *to the left* of $z = 0.67$ is 0.7486. Because we are asked the probability *greater* than $x = 170$ we need to have the area *to the right* of the curve, or

$$P(x > 170) = 1 - 0.7486 = 0.2514$$

or about 1/4. In other words, with a mean $\mu = 150$ and deviation $\sigma = 20$, we'd expect about a quarter of the time that the value of the variable would be greater than 170. Or, given our uncertainty of a specific value, we'd assign a probability of around 25% to it being larger than 170.

EXERCISE 7.1 *Given a Normal distribution, with parameters $\mu = 10$ and $\sigma = 2$, determine the following probabilities:*

1. $P(x < 12)$
2. $P(6 < x < 14)$
3. $P(2 < x < 12)$

EXERCISE 7.2 *Given a Normal distribution, with parameters $\mu = 2$ and $\sigma = 10$, answer the following questions (see Table 1.1 on page 45 for reference):*

1. *Make a qualitative plot of the distribution to help you with the other parts of the question*
2. *Is likely that $x > 0$?*
3. *Above which value of x is it very unlikely to observe?*
4. *Below which value of x is it extremely unlikely to observe?*

EXERCISE 7.3 *Given a Normal distribution, with parameters $\mu = 2$ and $\sigma = 0.5$, answer the following questions (see Table 1.1 on page 45 for reference):*

1. *Make a qualitative plot of the distribution to help you with the other parts of the question*
2. *Is likely that $x > 0$?*
3. *Above which value of x is it very unlikely to observe?*
4. *Below which value of x is it extremely unlikely to observe?*

Sum and Differences

One more convenient property of the Normal distribution is that sums and differences of variables that individually have Normal distributions also have Normal distributions, although each with a different mean and deviation parameter. The relationships are summarized as follows.

Sum of two Normally distributed variables If we have two variables, x and y, which have Normal distributions

$$\begin{aligned} P(x) &= \text{Normal}(\mu_x, \sigma_x) \\ P(y) &= \text{Normal}(\mu_y, \sigma_y) \end{aligned}$$

then their sum, $x + y$, has a mean the sum of the two, $\mu_x + \mu_y$ and a deviation $\sqrt{\sigma_x^2 + \sigma_y^2}$.

One way to remember this is that the new *squared* deviation parameter is the sum of the two old ones,

$$\sigma_{x+y}^2 = \sigma_x^2 + \sigma_y^2$$

Sum of two Normally distributed variables If we have two Normally distributed variables, x and y, we have

$$\begin{aligned} P(x) &= \text{Normal}(\mu_x, \sigma_x) \\ P(y) &= \text{Normal}(\mu_y, \sigma_y) \\ P(x+y) &= \text{Normal}(\mu_x + \mu_y, \\ &\quad \sqrt{\sigma_x^2 + \sigma_y^2}) \end{aligned}$$

Differences between two Normally distributed variables For differences, $x - y$, we have a new mean of $\mu_x - \mu_y$ and deviation parameter again $\sqrt{\sigma_x^2 + \sigma_y^2}$. Note the "+" sign in the new σ, which keeps the new σ positive which is must be by definition.

Differences between two Normally distributed variables

$$\begin{aligned} P(x-y) &= \text{Normal}(\mu_x - \mu_y, \\ &\quad \sqrt{\sigma_x^2 + \sigma_y^2}) \end{aligned}$$

(Note the "+" sign in the new σ.)

If we are asked for the distribution of a quantity with an added constant, like

$$z = x + \text{constant}$$

then the probability of z is just the same as that of x (i.e. Normal distribution with the same deviation), with the location parameter moved by the constant

$$P(z) = \text{Normal}(\mu_x + \text{constant}, \sigma_x)$$

EXAMPLE 7.3 *We have two Normal distributions* $P(x) = \text{Normal}(\mu = 8, \sigma = 2)$ *and* $P(y) = \text{Normal}(\mu = 20, \sigma = 7)$. *What is the distribution for* $z = y - x$?

The distribution $P(z)$ is also a Normal distribution, with mean $\mu_z = 20 - 8 = 12$ and deviation $\sigma_z = \sqrt{7^2 + 2^2} = 7.3$.

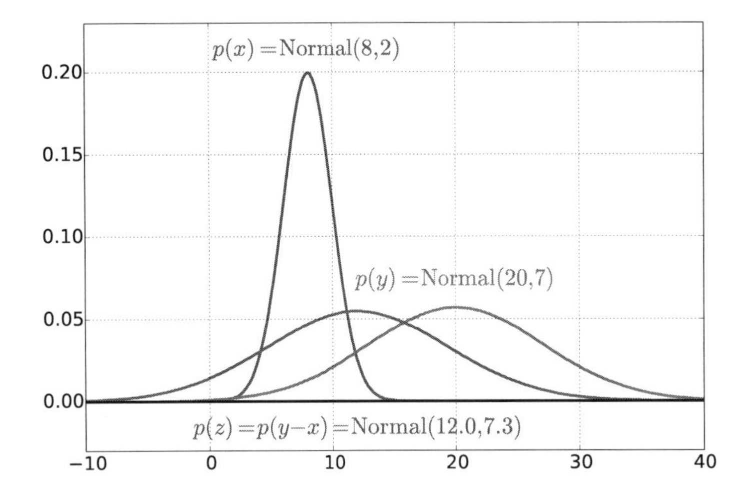

Significance

There is a term used in the literature called *statistical significance*.[2] Roughly it means a value that is *very unlikely* to be zero (see Table 1.1 on page 45), or in other words, the value of zero is not within the 95% percentile. This is within 2 standard deviations of the value, so the following estimated values are *not statistically significant*:

[2] Although the word "significant" occurs in the term "statistically significant", it does not imply that the result itself is important - it may be a small, uninteresting effect, but credibly non-zero. Perhaps a term like "statistically detectable" would be better, but we are unfortunately bound to the historical use of the term.

- 5 ± 3 - the *two*-deviation range is [-1,11] contains the value 0
- 7 ± 4
- -3 ± 2

but the following *are statistically significant*:

- 5 ± 2 - the two-deviation range is [1,9] does not contain the value 0
- 7 ± 3
- -3 ± 1

Statistical significance, at the *very unlikely* level (i.e. 95% percentile) is often used as a rough guideline to publish a positive effect.

An unintuitive consequence One consequence of this is that two studies that seem different may not be statistically different. The following example is from Gelman and Hill's book on Data Analysis.[3] Say, we have two measurements with means and standard deviations:

[3] A. Gelman, J. Hill, and Ebooks Corporation. *Data analysis using regression and multilevel/hierarchical models*, volume 625. Cambridge University Press Cambridge, UK:, 2007

- 25 ± 10
- 10 ± 10

number of deviations away from zero	term	probability
1σ	slightly likely/likely	0.7 (i.e. 7/10)
2σ	very likely	0.95 (i.e. 19/20)
3σ	extremely likely/virtually certain	0.002 (i.e. 1/500)
$> 4\sigma$	virtually certain	$> 999,999/1,000,000$

Table 7.1: Rough guide for the conversion of deviations away from zero and the qualitative labels for probability values for being a *significant* deviation.

The first study shows a significant effect (the two-deviation range is [5,45] does not contain zero), while the second one doesn't (the two-deviation range is [-10,30] contains zero). The *difference* between them is

$$(25 - 10) \pm \sqrt{10^2 + 10^2} = 15 \pm 14$$

which is *not* significant. One should be careful comparing the magnitudes and uncertainties of measurements!

7.3 The Normal Distribution - Estimating From Data

Estimating the mean, μ, knowing the deviation, σ

Typically one is provided with a series of measurements of a quantity, and we want to *estimate* the value of that quantity, and have a description of our *uncertainty* in the estimate. In Chapter 8 (*Applications of Parameter Estimation and Inference* on page 151) we go through a number of detailed examples of this process. Here, we simply summarize the result. We are given:

1. A series of N measurements, data=$\{x_1, x_2, x_3, \ldots, x_N\}$
2. The real deviation, σ
3. We are modeling the data as a true value, μ, with uncertainty with a likelihood from the Normal distribution with known deviation, σ, as in Normal$(0, \sigma)$. Further, we assume independence between the measurements.

Since in this case we are given σ, we wish then to estimate the parameter μ. The result will be a probability distribution over μ, with a best (i.e. most probable) value and an uncertainty in that value. The result is that the distribution of μ is also a Normal distribution,

$$P(\mu|\text{data}, \sigma) = \text{Normal}(\bar{x}, \sigma/\sqrt{N})$$

where the center value (and thus the most probable value of μ) is given by the sample mean of the data.

In scientific applications, this notation is often shortened to $\mu = \bar{x} \pm \sigma/\sqrt{N}$, so it is clear what is the best estimate of μ (i.e. $\bar{x}$) and what is the uncertainty in that estimate (i.e. $\sigma/\sqrt{N}$).

Sample Mean The *sample mean* of a set of N samples, $x_1, x_2, \cdots, x_N$ is given by

$$\bar{x} \equiv \frac{x_1 + x_2 + x_3 + \cdots + x_N}{N}$$

Sample Mean The *sample mean* of a set of N samples, $x_1, x_2, \cdots, x_N$ is given by

$$\bar{x} \equiv \frac{x_1 + x_2 + x_3 + \cdots + x_N}{N}$$

The uncertainty in μ is given by $\sigma/\sqrt{N}$. As a consequence, larger N (i.e. more data points), makes us more confident in the particular estimate for μ.

Estimate of location parameter μ given N samples and known deviation, σ In summary, the best estimate for the location parameter μ in the Normal distribution given a set of N samples, $x_1, x_2, \cdots, x_N$ is given by

$$\hat{\mu} = \frac{x_1 + x_2 + x_3 + \cdots + x_N}{N} \pm \sigma/\sqrt{N}$$

Estimate of location parameter μ given N samples and known deviation, σ The best estimate for the location parameter μ in the Normal distribution given a set of N samples, $x_1, x_2, \cdots, x_N$ is given by

$$\hat{\mu} = \frac{x_1 + x_2 + \cdots + x_N}{N} \pm \sigma/\sqrt{N}$$

Example 7.4 *Estimating the True Length of an Object*

Say we have an object, and 5 measurements of its length from the same ruler but from different people,

$$5.1[\text{cm}], 4.9[\text{cm}], 4.7[\text{cm}], 4.9[\text{cm}], 5.0[\text{cm}]$$

Say that we further know that the uncertainty (given this ruler) of one measurement has $\sigma = 0.5[\text{cm}]$. What is the best estimate of the length? The best estimate should be given by the sample mean of these 5 samples,

In real measurements, there is always the problem of bias or *systematic* uncertainties, where the uncertainty does not follow a Normal distribution. We will not consider this issue here.

$$\begin{aligned} \hat{\mu} &= \frac{x_1 + x_2 + \cdots + x_N}{N} \\ &= \frac{5.1[\text{cm}] + 4.9[\text{cm}] + 4.7[\text{cm}] + 4.9[\text{cm}] + 5.0[\text{cm}]}{5} = 4.92[\text{cm}] \end{aligned}$$

with uncertainty related to the known uncertainty of a single measurement,

$$\begin{aligned} \hat{\sigma} &= \frac{\sigma}{\sqrt{N}} \\ &= \frac{0.5[\text{cm}]}{\sqrt{5}} = 0.223[\text{cm}] \end{aligned}$$

yielding a final best estimate of

$$\hat{\mu} = 4.92[\text{cm}] \pm 0.223[\text{cm}]$$

or (with 2σ range),

$$\hat{\mu} = 4.92[\text{cm}], 95\% \text{ CI} = [4.474[\text{cm}], 5.366[\text{cm}]]$$

The 95% credible interval (CI) is really at the 1.96σ level, yielding $[4.481[\text{cm}], 5.358[\text{cm}]]$. We will almost always approximate it as 2σ by hand, but the computer will generate the true 95% credible interval when requested.

Estimating the mean, μ, not *knowing the deviation, σ*

If we are not so fortunate to be given the deviation, as in the previous case, then this parameter too must be estimated from the data. As a first step we can estimate the deviation with the *sample* deviation.

Sample Deviation The sample deviation of a set of N samples, $x_1, x_2, \cdots, x_N$ is given by

$$S \equiv \sqrt{\frac{1}{N-1}\left((x_1-\bar{x})^2+(x_2-\bar{x})^2+\cdots+(x_N-\bar{x})^2\right)}$$

Sample Deviation The sample deviation of a set of N samples, $x_1, x_2, \cdots, x_N$ is given by

$$S \equiv \sqrt{\frac{1}{N-1}\left((x_1-\bar{x})^2+\cdots+(x_N-\bar{x})^2\right)}$$

Approximate estimate of location parameter μ and deviation σ given N samples The posterior probability for μ and σ given a set of N samples, $x_1, x_2, \cdots, x_N$ can be approximated with

$$\begin{aligned} P(\mu|\text{data}) &\sim \text{Normal}(\bar{x}, S/\sqrt{N}) \\ P(\sigma|\text{data}) &\sim \text{Normal}\left(S, S^2/\sqrt{(N-1)/3}\right) \end{aligned}$$

which works well if we have many ($N > 30$) data points.

Approximate estimate of location parameter μ and deviation σ given N samples The posterior probability for μ and σ given a set of N samples, $x_1, x_2, \cdots, x_N$ can be approximated with

$$\begin{aligned} P(\mu|\text{data}) &\sim \text{Normal}(\bar{x}, S/\sqrt{N}) \\ P(\sigma|\text{data}) &\sim \text{Normal}\left(S, S^2/\sqrt{(N-1)/3}\right) \end{aligned}$$

which works well if we have many ($N > 30$) data points.

Because the uncertainty in the mean depends explicitly on the number of data points, it goes beyond the level of this chapter to give a form for the posterior probability distribution for the deviation, σ.

With a smaller data set, the value of S as an estimate for the deviation becomes too small. When the estimate for σ is too small, then the result is claiming *more confidence* in the estimate of the mean, μ, than is warranted. This discrepancy depends on the number of data points, and thus it makes sense that the proper distribution should depend on the number of data points, in addition to the sample mean and deviation. The proper, although less convenient, result is that the posterior probability for μ takes the form of the Student's t distribution,

Estimate of location parameter μ given N samples and *unknown* σ The posterior probability for μ takes the form of the Student's t distribution,

$$P(\mu|\text{data}) = \text{Student}_{\text{dof}=N-1}(\bar{x}, S/\sqrt{N})$$

This distribution requires *three* numbers to specify, referred to as the mean (μ), deviation (σ) and the *degrees of freedom* (dof). The degrees of freedom is defined in this case to be the number of data points less one, $N-1$.

Estimate of location parameter μ given N samples and *unknown* σ The posterior probability for μ takes the form of the Student's t distribution,

$$P(\mu|\text{data}) = \text{Student}_{\text{dof}=N-1}(\bar{x}, S/\sqrt{N})$$

This distribution requires *three* numbers to specify, referred to as the mean (μ), deviation (σ) and the *degrees of freedom* (dof). The degrees of freedom is defined in this case to be the number of data points less one, $N-1$.

EXAMPLE 7.5 *Estimating the True Length of an Object...Again*

Say we have an object, and 5 measurements of its length from the same ruler but from different people,

5.1[cm], 4.9[cm], 4.7[cm], 4.9[cm], 5.0[cm]

Unlike earlier, let's say that we don't know the uncertainty (given this ruler) of one measurement What is the best estimate of the length? Again, the best estimate should be given by the sample mean of these 5 samples,

$$\begin{aligned}
\hat{\mu} &= \frac{x_1 + x_2 + \cdots + x_N}{N} \\
&= \frac{5.1[\text{cm}] + 4.9[\text{cm}] + 4.7[\text{cm}] + 4.9[\text{cm}] + 5.0[\text{cm}]}{5} = 4.92[\text{cm}]
\end{aligned}$$

with uncertainty related to the sample deviation

$$\begin{aligned}
S^2 &= \frac{1}{N-1}\left((x_1 - \bar{x})^2 + \cdots + (x_N - \bar{x})^2\right) \\
&= \frac{1}{5-1}\left((5.1[\text{cm}] - 4.92[\text{cm}])^2 + (4.9[\text{cm}] - 4.92[\text{cm}])^2 + (4.7[\text{cm}] - 4.92[\text{cm}])^2 + \right. \\
&\quad \left. (4.9[\text{cm}] - 4.92[\text{cm}])^2 + (5.0[\text{cm}] - 4.92[\text{cm}])^2\right) \\
&= 0.024[\text{cm}]^2 \\
S &= \sqrt{0.024[\text{cm}]^2} = 0.155[\text{cm}] \\
\frac{S}{\sqrt{N}} &= \frac{0.155[\text{cm}]}{\sqrt{5}} = 0.069[\text{cm}]
\end{aligned}$$

Looking at Table D.2on page 194 with "Degrees of Freedom" equal to 4, we find that the 95% credible interval for μ (between areas 0.025 and 0.975) falls $\pm 2.776 \cdot S/\sqrt{N}$, thus we have

$$\begin{aligned}
\hat{\mu} &= 4.92[\text{cm}], 95\%\ \text{CI} = [4.92[\text{cm}] - 2.776 \cdot 0.069[\text{cm}], 4.92[\text{cm}] + 2.776 \cdot 0.069[\text{cm}]] \\
&= 4.92[\text{cm}], 95\%\ \text{CI} = [4.73[\text{cm}], 5.11[\text{cm}]]
\end{aligned}$$

Although much of this is easier with the computer, it is instructive to go through simple examples by hand.

7.4 *Normal Approximation*

The Normal distribution is useful for many reasons: its simple shape, the fact that there are only two parameters which describe it, and the ease with which one can compare the general Normal distribution to the single standard Normal. Further, it can be used as an approximation for several other distributions, under certain limits.

The Beta Distribution

We first saw the beta distribution as the posterior description in a bent-coin parameter estimation problem (see Section 6.3 on page 117 in Chapter 6 (*Introduction to Parameter Estimation*)). The Normal approximation occurs when the number of flips gets large, compared to how likely the coin flips heads. For notation, we will write the frequency of heads as

$$f \equiv \frac{h}{N}$$

Normal Approximation to the Beta Distribution The Normal Approximation to the Beta Distribution , for large number of flips (N) of which a fraction $f \equiv h/N$ are successful is given by

$$\text{Beta}(h, N) \sim \text{Normal}\left(\mu = f, \sigma = \sqrt{f(1-f)/N}\right)$$

Normal Approximation to the Beta Distribution The Normal Approximation to the Beta Distribution , for large number of flips (N) of which a fraction $f \equiv h/N$ are successful is given by

$$\text{Beta}(h, N) \sim \text{Normal}\left(\mu = f, \sigma = \sqrt{f(1-f)/N}\right)$$

To see how close this approximation can be, observe the following two cases:

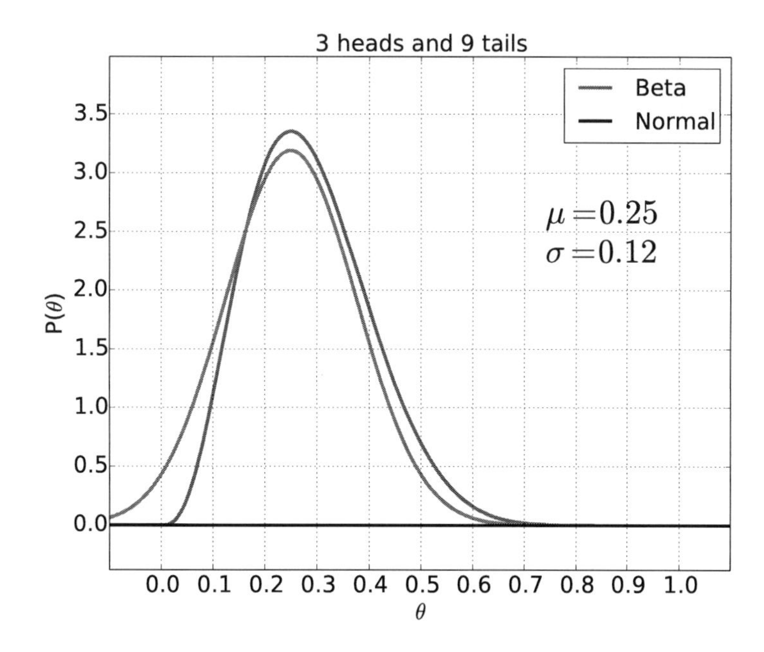

With ten times as many flips, we have

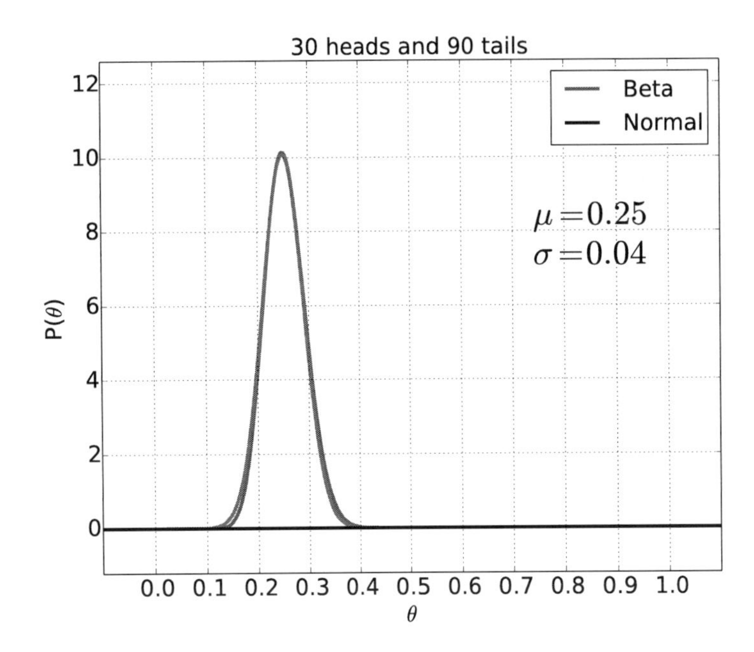

and the curves are so close as to be nearly identical! There still is a (small) probability for getting a negative θ, which is problematic in theory but not typically in practice. To use the properties of the Normal distribution here to quantify our uncertainty about the bent coin. Given 30 heads and 90 tails, the best estimate for θ (i.e. the top of the curve) is 0.25. Our uncertainty is quantified by the width of the distribution, given by σ. Thus, we can be confident to a 95% degree for θ within 2σ, or between 0.17 and 0.33 ($0.25 - 2 \cdot 0.04$ and $0.25 + 2 \cdot 0.04$, respectively).

This is an *approximation*, and as such will certainly give seriously incorrect answers under certain circumstances. For example, in this case, the Normal approximation predicts that there is around a 1.8% chance that the bent coin might have a *negative* θ, or probability of flipping heads (look a the Normal curve to the left of $\theta = 0$)! The beta distribution is zero for any value below zero or over one, and thus will never lead to such absurd answers.

The Binomial Distribution

Similarly, with the (discrete) binomial distribution (see Equation 3.3) we have the Normal approximation.

Normal Approximation to the Discrete Binomial Distribution

$$\text{Binomial}(N,p) = \text{Normal}(\mu = N \cdot p, \sigma = \sqrt{N \cdot p(1-p)})$$

with examples

Normal Approximation to the Discrete Binomial Distribution

$$\text{Binomial}(N,p) = \text{Normal}(\mu = N \cdot p, \sigma = \sqrt{N \cdot p(1-p)})$$

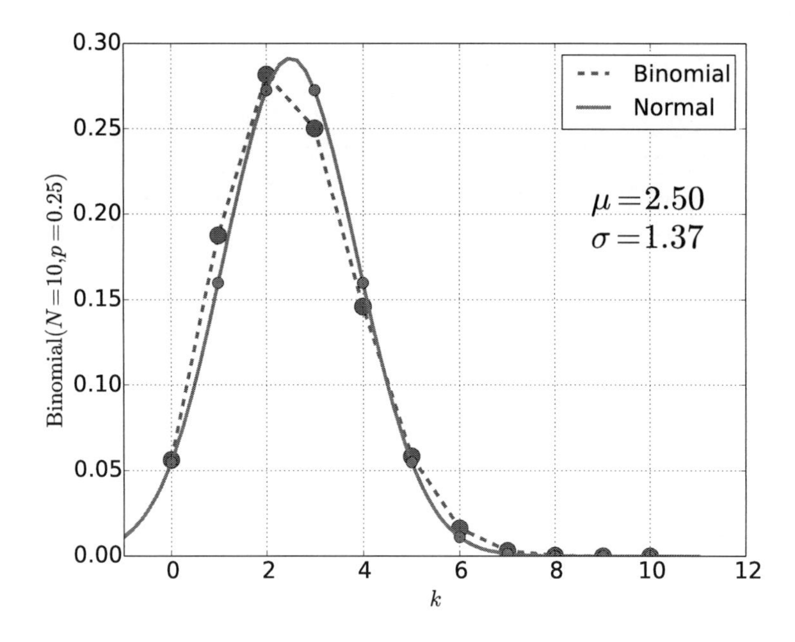

and

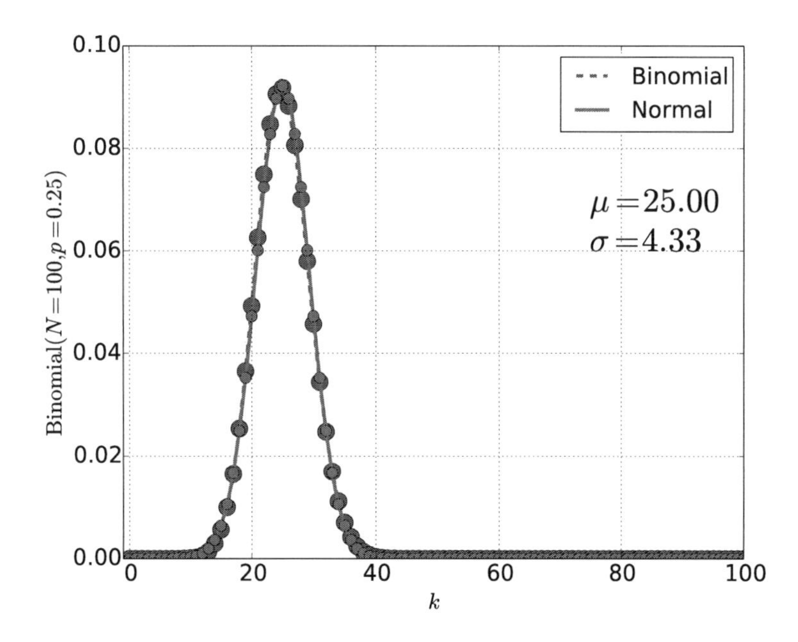

The Student's t Distribution

For smallish data sets, $5 < N < 30$, we can replace the estimate of the mean from the Student's t distribution to a Normal distribution with an increased estimate for the deviation. It then becomes practical to use the more convenient z-score to estimate credible intervals rather than the full t tables. The approximation in this domain looks like[4]

[4] D. Berry. *Statistics: A Bayesian Perspective.* Duxbury, 1996

Normal Approximation to the Student's t Distribution For smallish data sets, $5 < N < 30$,

$$\begin{aligned} \text{Student}_{\text{dof}=N-1}(\bar{x}, S/\sqrt{N}) &\sim \text{Normal}(\bar{x}, Sk/\sqrt{N}) \\ k &\equiv 1 + \frac{20}{N^2} \end{aligned}$$

Normal Approximation to the Student's t Distribution For smallish data sets, $5 < N < 30$,

$$\text{Student}_{\text{dof}=N-1}(\bar{x}, S/\sqrt{N}) \sim \text{Normal}(\bar{x}, k \cdot S/\sqrt{N})$$

$$k \equiv 1 + \frac{20}{N^2}$$

EXAMPLE 7.6 *Estimating the True Length of an Object...Yet Again*

Say we have an object, and 5 measurements of its length from the same ruler but from different people,

$$5.1[\text{cm}], 4.9[\text{cm}], 4.7[\text{cm}], 4.9[\text{cm}], 5.0[\text{cm}]$$

Unlike earlier, let's say that we don't know the uncertainty (given this ruler) of one measurement. What is the best estimate of the length? Again, the best estimate should be given by the sample mean of these 5 samples,

$$\begin{aligned} \hat{\mu} &= \frac{x_1 + x_2 + \cdots + x_N}{N} \\ &= \frac{5.1[\text{cm}] + 4.9[\text{cm}] + 4.7[\text{cm}] + 4.9[\text{cm}] + 5.0[\text{cm}]}{5} = 4.92[\text{cm}] \end{aligned}$$

with uncertainty related to the adjusted sample deviation,

$$\begin{aligned} S^2 &= \frac{1}{N-1}\left((x_1 - \bar{x})^2 + \cdots + (x_N - \bar{x})^2\right) \\ &= \frac{1}{5-1}\Big((5.1[\text{cm}] - 4.92[\text{cm}])^2 + (4.9[\text{cm}] - 4.92[\text{cm}])^2 + (4.7[\text{cm}] - 4.92[\text{cm}])^2 + \\ &\quad (4.9[\text{cm}] - 4.92[\text{cm}])^2 + (5.0[\text{cm}] - 4.92[\text{cm}])^2\Big) \\ &= 0.024[\text{cm}]^2 \\ S &= \sqrt{0.024[\text{cm}]^2} = 0.155[\text{cm}] \\ \frac{S}{\sqrt{N}} &= \frac{0.155[\text{cm}]}{\sqrt{5}} = 0.069[\text{cm}] \\ k &= 1 + \frac{20}{5^2} = 1.8 \\ k \cdot \frac{S}{\sqrt{N}} &= 1.8 \cdot 0.069[\text{cm}] = 0.124[\text{cm}] \end{aligned}$$

yielding a final best estimate of

$$\hat{\mu} = 4.92[\text{cm}] \pm 0.124[\text{cm}]$$

or (with 2σ range),

$$4.92[\text{cm}], 95\% \text{ CI} = [4.672[\text{cm}], 5.168[\text{cm}]]$$

Compare this range to the one shown in Example 7.5 on page 141. The one here has a slightly larger range, which is a bit more conservative than is needed, but the calculation is quite a bit easier.

7.5 Summary

It is useful to see all of these results stemming from the same Bayes' Recipe, applied to different models of the data and (possibly) different prior probabilities. As we have stated, many of the simple cases have been worked out by the mathematicians, so we don't need to do the work of deriving them. It will be our task to understand their properties, to be able to apply them to real problems, and to understand their consequences. One of the immediate observations that we make is the prevalence of the *Normal* distribution, justifying our detailed exploration of it in this chapter.

1 ***Proportions***

Parameter of Interest: θ, the chances of a single event

Applications: coin flips, voting percentages, success in sports, performance on tests

Form of the data: h successes in N total events

Model of the data:

$$\text{data} = \begin{cases} \text{success} & \text{, with probability } \theta \\ \text{failure} & \text{, otherwise (i.e. with probability } 1-\theta) \end{cases}$$

Posterior Probability:

$$\underbrace{\text{Beta}(\theta|\text{data})}_{\text{posterior probability}} \sim \overbrace{\text{Binomial}(\text{data}|\theta)}^{\text{likelihood}} \times \underbrace{\text{Uniform}(\theta)}_{\text{prior probability}}$$

2 ***Magnitude with Known Deviation***

Parameter of Interest: μ, the true magnitude of a quantity, given the deviation, labeled by σ, from the central value

Applications: percentages with large samples, scientific measurements such as weight and size of objects, time scales of events

Form of the data: N total data points, labeled $x_1, x_2, \cdots, x_N$, and given known σ

Model of the data:

$$\text{data} = \mu + \text{uncertainty with probability Normal}(\mu = 0, \text{known } \sigma)$$

Posterior Probability:

$$\underbrace{\text{Normal}(\mu_2|\text{data}, \sigma)}_{\text{posterior probability}} \sim \overbrace{\text{Normal}(\text{data}|\mu, \sigma)}^{\text{likelihood}} \times \underbrace{\text{Uniform}(\mu)}_{\text{prior probability}}$$

3 *Magnitude with Unknown Deviation*

Parameter of Interest: μ, the true magnitude of a quantity, and the unknown deviation, labeled by σ, from the central value

Applications: scientific measurements with small samples (less than around 30), such as weight and size of objects, time scales of a small number of events

Form of the data: N total data points, labeled $x_1, x_2, \cdots, x_N$

Model of the data:

$$\text{data} = \mu + \text{uncertainty with probability Normal}(\mu = 0, \sigma)$$

Posterior Probability:

$$\underbrace{\text{P}(\mu, \sigma|\text{data})}_{\text{posterior probability}} \sim \overbrace{\text{Normal}(\text{data}|\mu, \sigma)}^{\text{likelihood}} \times \underbrace{\text{Uniform}(\mu) \cdot \text{Uniform}(\log \sigma)}_{\text{prior probability}}$$

$$\underbrace{\text{Student} - \text{T}(\mu|\text{data})}_{\text{posterior probability}} \sim [\text{P}(\mu, \sigma|\text{data})]_{\text{marginalized over } \sigma}$$

$$\underbrace{\text{F}(\sigma|\text{data})}_{\text{posterior probability}} \sim [\text{P}(\mu, \sigma|\text{data})]_{\text{marginalized over } \mu}$$

7.6 *Computer Examples*

```
from sie import *
```

Estimating Lengths

Known deviation, σ

```
x=[5.1, 4.9, 4.7, 4.9, 5.0]
sigma=0.5
```

```
mu=sample_mean(x)
N=len(x)
```

```
dist=normal(mu, sigma/sqrt(N))
distplot(dist)
```

```
<matplotlib.figure.Figure at 0x10713c710>
```

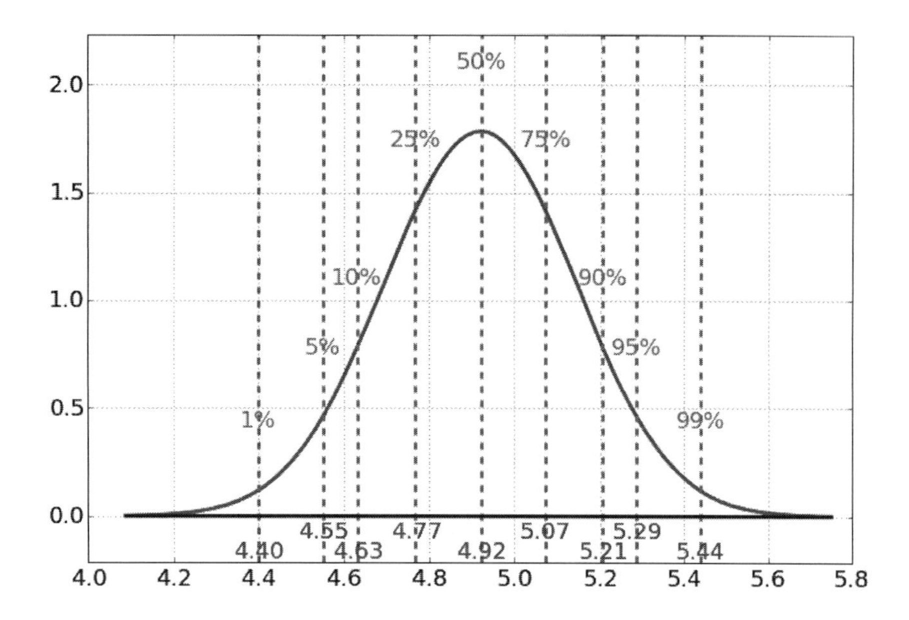


```
credible_interval(dist)
```

```
(4.4817387297117088, 4.9199999999999999, 5.358261270288291)
```

Unknown σ

```
mu=sample_mean(x)
s=sample_deviation(x)
print mu,s
```

```
4.92 0.148323969742
```

```
dist=tdist(N-1,mu,s/sqrt(N))
```

```
distplot(dist,xlim=[4.6,5.4])
```

```
<matplotlib.figure.Figure at 0x1085b5c50>
```

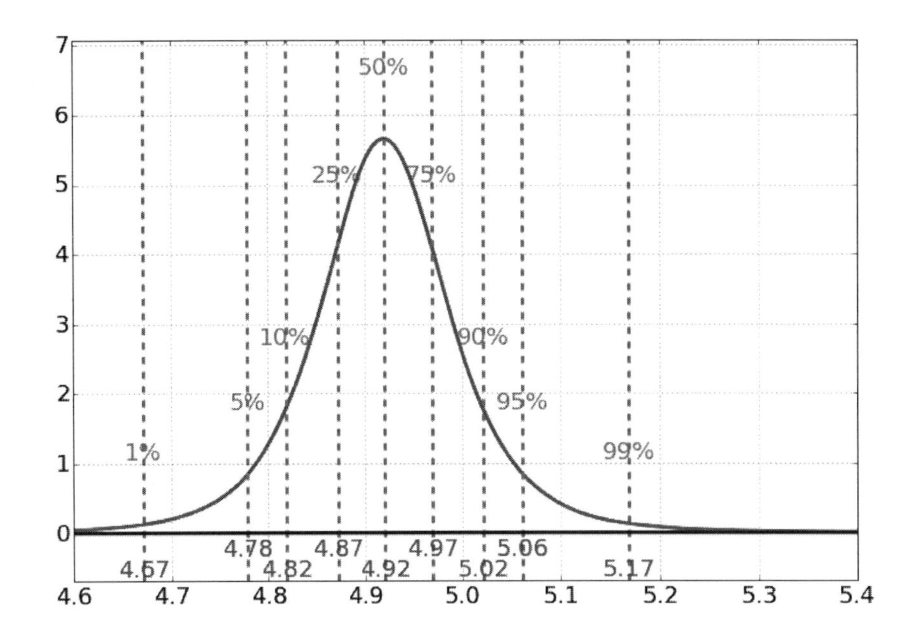


```
credible_interval(dist)
```

```
(4.7358314667008017, 4.9199999999999999, 5.1041685332991982)
```

8 Applications of Parameter Estimation and Inference

8.1 Normal Model - Inference about Means

Example 8.1 *Iris petal lengths - Best estimate*

1.4	1.4	1.3	1.5	1.4
1.7	1.4	1.5	1.4	1.5
1.5	1.6	1.4	1.1	1.2
1.5	1.3	1.4	1.7	1.5
1.7	1.5	1.0	1.7	1.9
1.6	1.6	1.5	1.4	1.6
1.6	1.5	1.5	1.4	1.5
1.2	1.3	1.5	1.3	1.5
1.3	1.3	1.3	1.6	1.9
1.4	1.6	1.4	1.5	1.4

Table 8.1: Iris petal lengths, in centimeters, for Iris type *Setosa*.

Table 8.1 shows data for the lengths (in centimeters) of the petals of one species of Iris flower[1]. If we want to estimate the "true" length of the the petal for this species, given all of these examples, we would apply the following model of the data:

[1] K. Bache and M. Lichman. UCI machine learning repository, 2013. URL `http://archive.ics.uci.edu/ml`

$$\text{data} = \text{true value} + \text{Normal(mean=0,known } \sigma)$$

or equivalently

$$\text{data} = \text{Normal(mean=true value,known } \sigma)$$

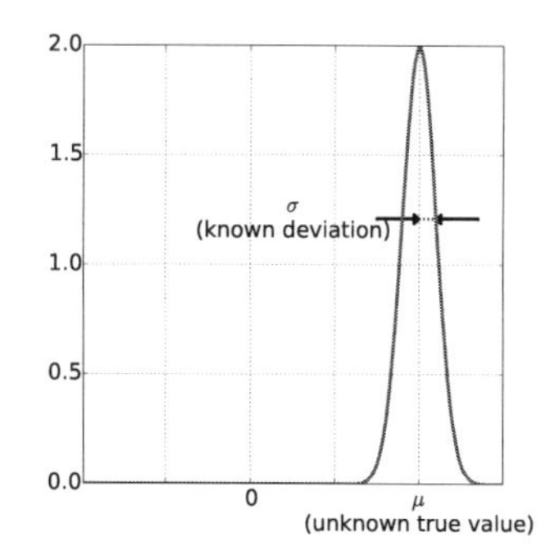

The resulting distribution for the "true value", μ, is also a *Normal* distribution (Section 7.3),

$$P(\mu|\text{data}, \sigma) = \text{Normal}(\bar{x}, \sigma/\sqrt{N})$$

where the best estimate of the true value, μ is the sample mean, $\bar{x}$, and the uncertainty is related to the sample deviation (which we're going to take as the "known" deviation, $\sigma \sim 0.174$) in this case. Thus,

$$\begin{aligned}\hat{\mu} &= \bar{x} = \frac{1.4 + 1.4 + 1.3 + 1.5 + \cdots + 1.6 + 1.4 + 1.5 + 1.4}{50} \\ &= 1.464\end{aligned}$$

and the full answer, with uncertainty, is

$$\begin{aligned}\hat{\mu} &= 1.464[\text{cm}] \pm \frac{0.174}{\sqrt{50}}[\text{cm}] \\ &= 1.464[\text{cm}] \pm 0.025[\text{cm}]\end{aligned}$$

Example 8.2 *Iris petal lengths - A different species?*

Imagine we have a single observation of another iris with petal length 2.5 [cm]. Is this likely to be the same type as the *Setosa* type above? As outlined in Section 7.2, we get the best estimate for the difference as:

$$\mu_{\text{diff}} = 2.5 - 1.464 = 1.036$$

with uncertainty the same as the uncertainty of the *Setosa* type, so the final estimate with uncertainty is:

$$1.036[\text{cm}] \pm 0.025[\text{cm}]$$

which is

$$\frac{1.036[\text{cm}]}{0.025[\text{cm}]} = 41 \text{ deviations away from zero!}$$

which makes it *virtually certain* to be a different type (see Table 7.1).

8.2 *Normal Model Again - Inference about Means and Deviations*

Setosa	1.4	1.4	1.3	1.5	1.4
Virginica	6.0	5.1	5.9	5.6	5.8
Versicolor	4.7	4.5	4.9	4.0	4.6

Table 8.2: Subset of iris petal lengths, in centimeters, for iris types *Virginica*, *Setosa*, and *Versicolor*.

Example 8.3 *Iris petal lengths - Significantly different?*

Shown in Table 8.2 is a very small subset of the full iris petal-length data. Are the types *Virginica* and *Versicolor* longer than the type *Setosa*? Is the *Virginica* longer than *Versicolor*? For each of these, we need to specify the model, determine the best estimate for the parameters of the model, and then compare the distributions.

The model we will use is the simple Normal model,

$$\text{data} = \text{Normal(mean=true value,unknown } \sigma)$$

which is the same as the previous example, except that the deviation, σ, is unknown. In addition to being unknown, there are so few data points that the deviation can't be well approximated with the sample deviation.

The resulting distribution for the "true value", μ, is a *Student-t* distribution (Section 7.3),

$$P(\mu|\text{data}) = \text{Student}_{\text{dof}=N-1}(\bar{x}, S/\sqrt{N})$$

The best estimates for the true length-values of each type is given by their sample means,

$$\begin{aligned}
\hat{\mu}_{\text{setosa}} &= \frac{1.4+1.4+1.3+1.5+1.4}{5} = 1.40 \\
\hat{\mu}_{\text{virginica}} &= \frac{6.0+5.1+5.9+5.6+5.8}{5} = 5.68 \\
\hat{\mu}_{\text{versicolor}} &= \frac{4.7+4.5+4.9+4.0+4.6}{5} = 4.54
\end{aligned}$$

and the sample deviations for each is given by

$$\begin{aligned}
S_{\text{setosa}} &= \sqrt{\frac{1}{5-1}\cdot((1.4-1.40)^2+(1.4-1.40)^2+(1.3-1.40)^2+(1.5-1.40)^2+(1.4-1.40)^2)} \\
&= = 0.07 \\
S_{\text{virginica}} &= \sqrt{\frac{1}{5-1}\cdot((6.0-5.68)^2+(5.1-5.68)^2+(5.9-5.68)^2+(5.6-5.68)^2+(5.8-5.68)^2)} \\
&= = 0.36 \\
S_{\text{versicolor}} &= \sqrt{\frac{1}{5-1}\cdot((4.7-4.54)^2+(4.5-4.54)^2+(4.9-4.54)^2+(4.0-4.54)^2+(4.6-4.54)^2)} \\
&= 0.34
\end{aligned}$$

The posterior probability distributions, shown in Figure 8.1, have the following form:

$$\begin{aligned}
P(\mu_{\text{setosa}}|\text{data}) &= \text{Student}_{\text{dof}=4}(1.40, 0.07/\sqrt{5}) \\
P(\mu_{\text{virginica}}|\text{data}) &= \text{Student}_{\text{dof}=4}(5.68, 0.36/\sqrt{5}) \\
P(\mu_{\text{versicolor}}|\text{data}) &= \text{Student}_{\text{dof}=4}(4.64, 0.34/\sqrt{5})
\end{aligned}$$

It is clear from the picture that they are very well separated, but we can quantify this by looking at the probability that the difference between their means is greater than zero.

The probability of their difference approximately takes the form of a Student's t distribution, with the same center and deviation shown for the Normal in Section 7.2. Here we do the calculation between the closest two iris types, *Virginica* and *Versicolor*:

This approximation is called Welch's method. The exact analysis is beyond this book, but numerically one can calculate it and it doesn't differ from this approximate analysis in any significant way. Essentially you calculate $P(\mu_{\text{versicolor}} > \mu_{\text{virginica}}|\text{data})$ by adding up the $P(\mu_{\text{versicolor}}|\text{data}) \times P(\mu_{\text{virginica}}|\text{data})$ for all possible lengths where versicolor is longer than virginica.

$$\begin{aligned}
\mu_{\text{diff}} &= 5.68 - 4.64 = 1.04 \\
\sigma_{\text{diff}} &= \sqrt{\frac{0.36^2}{5} + \frac{0.34^2}{5}} = 0.22
\end{aligned}$$

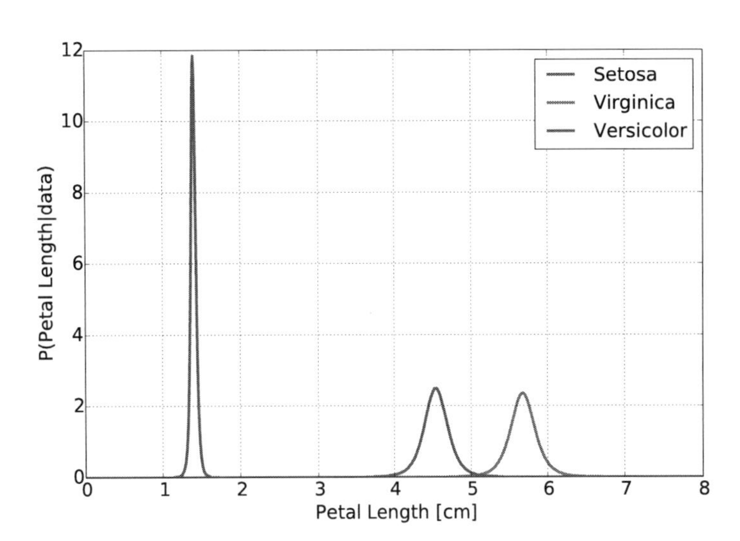

Figure 8.1: Probability distributions for the subset of iris petal lengths. Each distribution follows a Student-t form.

The degrees of freedom used for this Student's t distribution is approximately the smallest one from the two samples, or in this case (since both samples have the same number of data points), dof=4. The resulting posterior probability distribution for the difference of means is shown in Figure 8.2.

We observe that the difference of the means is over 4 times the deviation away from zero, so even with 4 degrees of freedom, this is significant at the 99% level. We can be highly certain that these two species have different petal lengths, and that the difference observed is not just a product of the random sample.

EXAMPLE 8.4 *Ball Bearing Sizes*

Here's a data data set, measuring the size of ball bearings[2] from two different production lines.

First line [microns]									
1.18	1.42	0.69	0.88	1.62	1.09	1.53	1.02	1.19	1.32
Second line [microns]									
1.72	1.62	1.69	0.79	1.79	0.77	1.44	1.29	1.96	0.99

Table 8.3: Production lines are produce a ball bearing with a diameter of approximately 1 micron. Ten ball bearings were randomly picked from the production line (i.e. the *First line*) at one time, and then again for a different production line (i.e. the *Second line*). Romano, A. (1977) *Applied Statistics for Science and Industry.*

We can ask questions such as:

- What is our best estimate of the size of a ball bearing, given one of the production lines?
- Is it reasonable to believe that there is a difference in the size produced between the two lines?

[2] David J Hand, Fergus Daly, K McConway, D Lunn, and E Ostrowski. *A handbook of small data sets*, volume 1. CRC Press, 2011

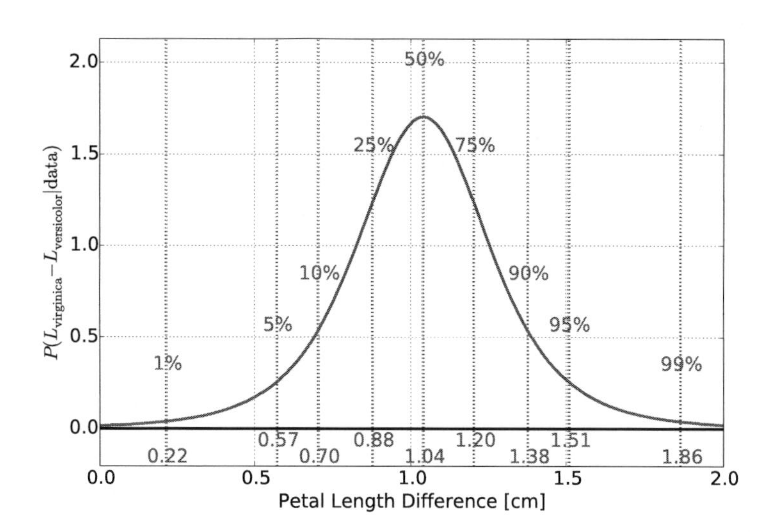

Figure 8.2: Probability distributions for the difference between iris petal lengths for the closest two iris types, *Virginica* and *Versicolor*. The distribution follows a Student-t form, and clearly shows significant probability (greater than 99%) for being greater than zero.

EXAMPLE 8.5 *What is the best estimate (and uncertainty) for each of the two production lines of ball bearings?*

Using the normal approximation to the Student-T distribution (Section 7.4), we have the best estimates of the two lines as

$$\begin{aligned}\mu_1 &= \frac{1.180000 + 1.420000 + 0.690000 + \cdots + 1.190000 + 1.320000}{10} \\ &= 1.194\end{aligned}$$

$$\begin{aligned}\mu_2 &= \frac{1.720000 + 1.620000 + 1.690000 + \cdots + 1.960000 + 0.990000}{10} \\ &= 1.406\end{aligned}$$

and their uncertainties calculated by first calculating the sample deviations

$$\begin{aligned}S_1 &= \sqrt{\frac{1}{10-1} \cdot ((1.18 - 1.194)^2 + (1.42 - 1.194)^2 + \cdots + (1.19 - 1.194)^2 + (1.32 - 1.194)^2)} \\ &= 0.289 \\ S_2 &= \sqrt{\frac{1}{10-1} \cdot ((1.72 - 1.406)^2 + (1.62 - 1.406)^2 + \cdots + (1.96 - 1.406)^2 + (0.99 - 1.406)^2)} \\ &= 0.428\end{aligned}$$

and then scaling the deviations by the number of data points

$$\begin{aligned}\sigma_1 &= S_1/sqrt10 = 0.092 \\ \sigma_2 &= S_2/sqrt10 = 0.135\end{aligned}$$

yielding the best estimates and uncertainties for the two production lines

- Production line 1: 1.194 [microns]$\pm$ 0.092[microns]
- Production line 2: 1.406 [microns]$\pm$ 0.135[microns]

or looking at the 95% CI for each line

This is just the $\pm 2 \cdot \sigma$ range

- Production line 1: 1.01[microns] - 1.378[microns]
- Production line 2: 1.136[microns] - 1.676[microns]

Roughly, given that these intervals overlap, there is not strong evidence that there is a difference between the two lines.

EXAMPLE 8.6 *Is it reasonable to believe that there is a difference in the size produced between the two lines?*

Using the best estimate of the difference, we get

$$\delta_{12} = \mu_2 - \mu_1 = 0.212$$

with the uncertainty in the difference from the individual uncertainties,

$$\sigma_{12} = \sqrt{\sigma_1^2 + \sigma_2^2} = 0.163$$

So the $2 \cdot \sigma$ uncertainty range for the difference,

$$[0.212 - 2 \cdot 0.163, 0.212 + 2 \cdot 0.163] = [-0.114, 0.538]$$

includes the value zero, which we interpret as a statement that the difference is *not statistically significant*. In other words, it is *not reasonable* to believe that there is a difference in the size produced between the two lines.

8.3 *Beta Model - Inference About Proportions*

EXAMPLE 8.7 *The Sunrise Problem*

The sunrise problem, as first stated by Laplace, is "What is the probability that the sun will rise tomorrow?" We'll start with the assumption that initially one has never seen a sunrise, and then observe a year of sunrises each morning with no morning without one. Thus we have the form of the data as h successes (days with a sunrise) in N total days. Our model of the data is specified as before with a binomial distribution, resulting in the posterior Beta, as described in Section 6.6.

After a only 10 years of watching sunrises, and no failures of a sunrise, the best estimate for the probability of a sunrise is

$$\begin{aligned} \hat{\theta}_{\text{median}} &\approx \frac{h+2}{N+4} \\ &= \frac{3650+2}{3650+4} = 0.9995 \end{aligned}$$

making it virtually certain for a sunrise.

EXAMPLE 8.8 *Cancer Rates*

This example is from Donald Berry's Statistics textbook[3]:

[3] D. Berry. *Statistics: A Bayesian Perspective.* Duxbury, 1996

> pp 192: A study (Murphy and Abbey, Cancer in Families, 1959) addressed the question of whether cancer runs in families. The investigator identified 200 women with breast cancer and another 200 women without breast cancer and asked them whether their mothers had had breast cancer. Of the 400 women in the two groups combined, 10 of the mothers had had breast cancer. If there is no genetic connection, then about half of these 10 would come from each group.

The data is that 7 of the daughters had cancer and 3 did not. Is there strong evidence of a connection?

The proper way, assuming total initial ignorance, is to use the Beta distribution:

$$P(\theta_{\text{cancer}}|\text{data}) = \text{Beta}(h = 7, N = 10)$$

which has a median of $\hat{\theta}_{\text{cancer}} = 0.68$, but a 95% credible interval of $\hat{\theta}_{\text{cancer}} = 0.39$ up to $\hat{\theta}_{\text{cancer}} = 0.89$. This means there is not strong evidence of an effect.

EXAMPLE 8.9 *Cancer Rates - Normal Approximation*

We can estimate the the Beta distribution median and credible intervals with a Normal distribution, by using the "assuming 2 successes and 2 failures" method.

$$\begin{aligned} \hat{\theta}_{\text{cancer}} &= \frac{h+2}{N+4} \\ &= \frac{7+2}{10+4} = 0.643 \end{aligned}$$

and

$$\begin{aligned} \sigma &= \sqrt{\hat{\theta}_{\text{cancer}}(1-\hat{\theta}_{\text{cancer}})/(N+4)} \\ &= \sqrt{0.643(1-0.643)/(10+4)} \\ &= 0.128 \end{aligned}$$

So the approximate 95% credible interval is

$$\hat{\theta}_{\text{cancer}} \pm 2\sigma$$

which is between 0.387 and 0.899, again with the same conclusion of no strong evidence of an effect.

EXAMPLE 8.10 *Will it rain on the 4th of July?*

In the United States, the 4th of July is Independence Day, and is known for parades. The oldest continuously running parade is in Bristol, RI, and it runs rain or shine. Is it likely to rain on the parade? Climate data from nearby Providence is here from wunderground.com:

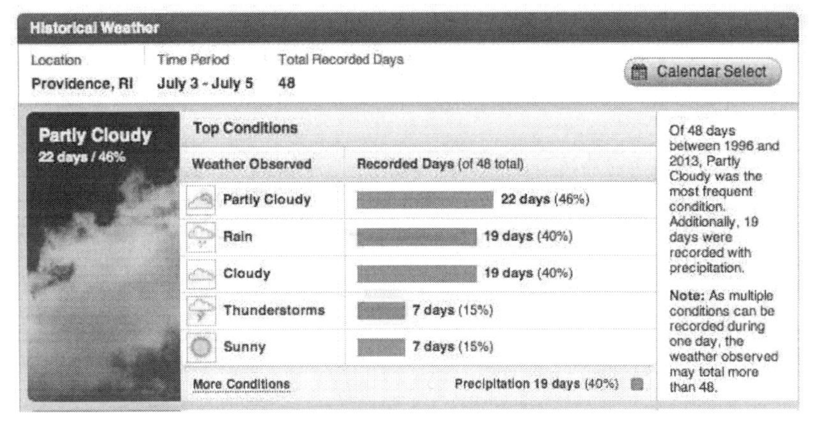

We can estimate the the Beta distribution median and credible intervals with a Normal distribution, by using the "assuming 2 successes and 2 failures" method.

$$\begin{aligned}\hat{\theta}_{\text{rain}} &= \frac{h+2}{N+4} \\ &= \frac{19+2}{48+4} = 0.404\end{aligned}$$

around 40%, less than an even chance (50%) of rain, but

$$\begin{aligned}\sigma &= \sqrt{\hat{\theta}_{\text{rain}}(1-\hat{\theta}_{\text{rain}})/(N+4)} \\ &= \sqrt{0.404(1-0.404)/(48+4)} \\ &= 0.068\end{aligned}$$

So the approximate 95% credible interval is

$$\hat{\theta}_{\text{rain}} \pm 2\sigma$$

which is between 0.268 and 0.540. This is not strong evidence against a purely fair and random "coin flip" for rain on the 4th of July.

EXAMPLE 8.11 *Hot Hand Reexamined*

In Tversky and Gilovich[4] we have the following data for Larry Bird free throws in basketball:

- Given each of 53 missed shots, Larry Bird successfully shot 48 of the *next* attempt.
- Given each of 285 successful shots, Larry Bird successfully shot 251 of the *next* attempt.

[4] A. Tversky and T. Gilovich. The cold facts about the" hot hand" in basketball. *Anthology of statistics in sports*, 16:169, 2005

This data alone almost suggests an anti-hot-hand (where you're *less* likely to make a successful attempt following a successful shot). However, we can demonstrate that these numbers are not in fact statistically different. Given the relatively large number of attempts (greater than 30) we can use the Normal approximation to estimate the two probabilities of success:

$$\theta_{\text{after a miss}} = \frac{48+2}{53+4} = 0.877$$
$$\theta_{\text{after a success}} = \frac{251+2}{285+4} = 0.875$$

and the uncertainty,

$$\sigma_{\text{after a miss}} = \sqrt{0.877(1-0.877)/(53+4)} = 0.044$$
$$\sigma_{\text{after a success}} = \sqrt{0.875(1-0.875)/(285+4)} = 0.019$$

making the 95% credible intervals for probability of a Larry Bird successful attempt

$$95\%\text{CI}_{\text{after a miss}} = 0.877 \pm 2 \cdot 0.044 = [0.789, 0.965]$$
$$95\%\text{CI}_{\text{after a success}} = 0.875 \pm 2 \cdot 0.019 = [0.837, 0.913]$$

Notice that the intervals overlap, so there is no significant evidence for a difference in Larry Bird's success following another success or following a miss. Thus, there is no significant evidence for a hot hand, or an anti-hot hand.

8.4 Model Construction

In practice, we either don't know what the optimum model we need is, or the needs of the model change as we obtain more data.

We start with the data in Table 8.4 for the mass of pennies of various years (shown graphically in Figure 8.3)[5]:

We are going to ignore the measurement uncertainties in these individual measurements, because they are quite small.

[5] this data was extracted from student measurements during a physics lab

EXAMPLE 8.12 *Mass of the Penny, Model 1 - One True Value*

If we assume a model that states that there is a "true" value and the variation from this "true" value caused by some unknown process, but with known magnitude, σ,

$$\text{data} = \text{true value} + \text{Normal(mean=0,known } \sigma)$$

or equivalently

$$\text{data} = \text{Normal(mean=true value,known } \sigma)$$

Year	Mass [g]
1960	3.133
1961	3.083
1962	3.175
1963	3.120
1964	3.100
1965	3.060
1966	3.100
1967	3.100
1968	3.073
1969	3.076
1970	3.100
1971	3.110
1972	3.080
1973	3.100
1974	3.093

Table 8.4: Mass of Pennies from 1960 to 1974.

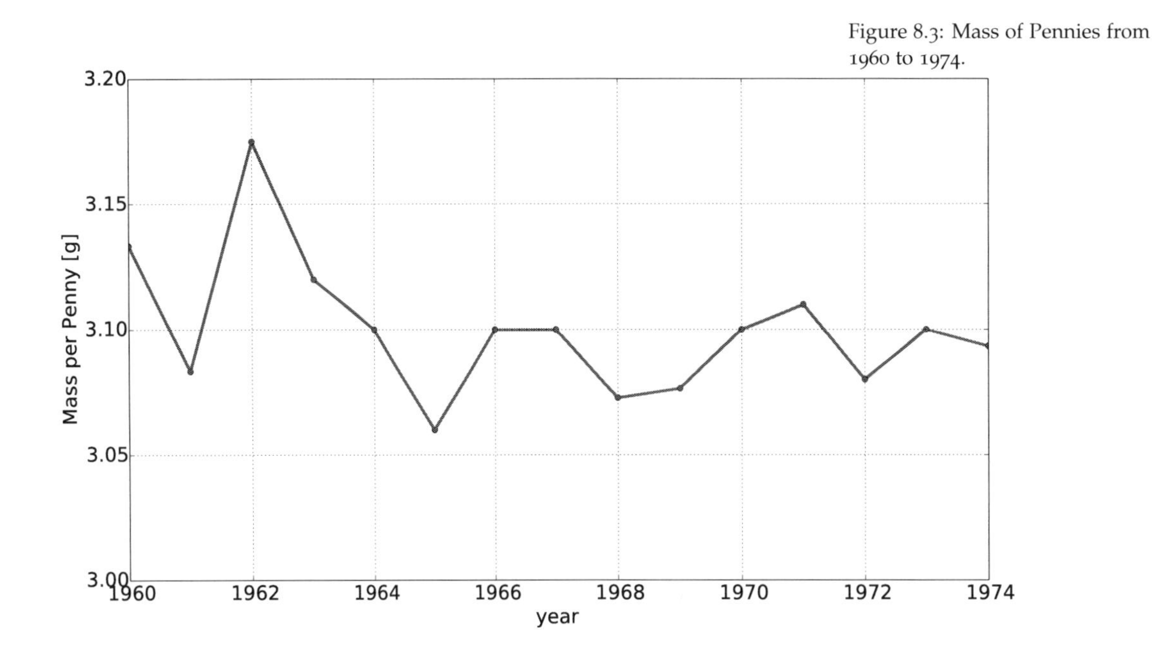

Figure 8.3: Mass of Pennies from 1960 to 1974.

we can get the best estimate and uncertainty in that estimate from the following procedure, Using the normal approximation to the Student-T distribution (Section 7.4):

$$\hat{\mu} = \bar{x} \pm k \cdot S/\sqrt{N}$$

where the symbols in this equation are

1 the number of data points, N.

2 the best estimate for the true value, $\hat{\mu}$, is given by the *sample mean*, $\bar{x}$:

$$\begin{aligned} \bar{x} &= = \frac{x_1 + x_2 + \cdots + x_N}{N} \\ &= \frac{3.133\text{g} + 3.083\text{g} + \cdots + 3.093\text{g}}{15} \\ &= 3.100\text{g} \end{aligned}$$

3 The uncertainty is directly related to the *sample standard deviation*, S:

$$\begin{aligned} S &= \sqrt{\frac{(x_1 - \bar{x})^2 + (x_2 - \bar{x})^2 + \cdots + (x_N - \bar{x})^2}{N-1}} \\ &= \sqrt{\frac{(3.133\text{g} - 3.100\text{g})^2 + (3.083 - 3.100\text{g})^2 + \cdots + (3.093 - 3.100\text{g})^2}{14}} \\ &= 0.0278\text{g} \end{aligned}$$

4 The scale factor, k, adjusts for the small number of data points - there is more uncertainty in our estimate when there are fewer data points:

$$\begin{aligned} k &= 1 + \frac{20}{N^2} \\ &= 1 + \frac{20}{15^2} \\ &= 1.0889 \end{aligned}$$

Finally, we have the best estimate and uncertainty for the pennies in this dataset:

$$\begin{aligned} \hat{\mu} &= \bar{x} \pm k \cdot S/\sqrt{N} \\ &= 3.100\text{g} \pm 1.0889 \cdot 0.0278\text{g}/\sqrt{15} \\ &= 3.100\text{g} \pm 0.0078\text{g} \end{aligned}$$

or, as a 99% credible range (3 times the uncertainty written above), we have, (see also Figure 8.4)

$$\begin{aligned} 99\% \text{ CI for } \mu &= 3.100\text{g} \pm 3 \times 0.0078\text{g} \\ &= [3.077\text{g}, 3.124\text{g}] \end{aligned}$$

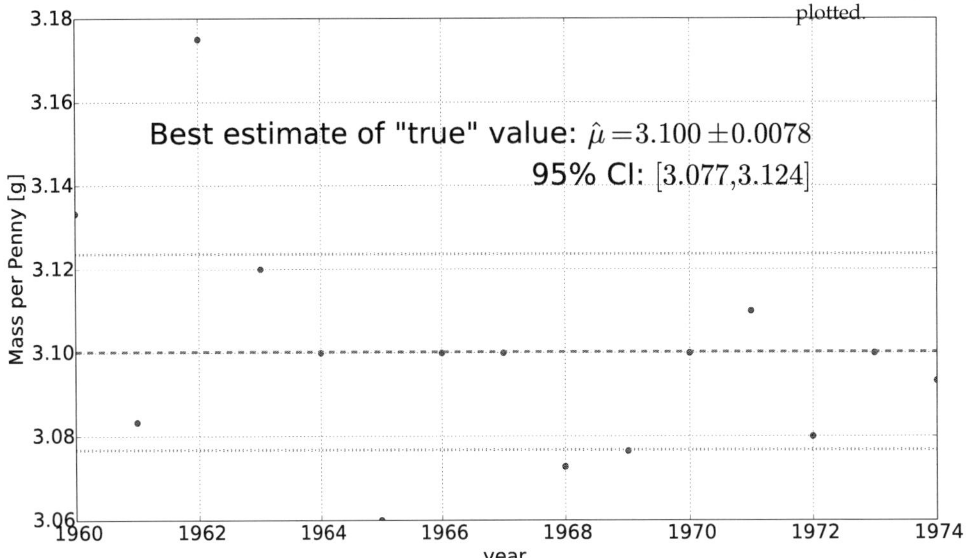

Figure 8.4: Mass of Pennies from 1960 to 1974, with best estimates and 99% CI (i.e. 3σ) uncertainty plotted.

EXAMPLE 8.13 *Mass of the Penny, Model 1 - One True Value with More Data*

Now we collect the additional data with more recent pennies shown in Table 8.5. We can follow the same procedure, assuming our original model of one "true" value, to get the best estimate and uncertainty for this model, combining the two data sets.

$$\hat{\mu} = \bar{x} \pm k \cdot S/\sqrt{N}$$

where the symbols in this equation are

1 the number of data points, $N = 30$.

2 the best estimate for the true value, $\hat{\mu}$, is given by the *sample mean*, $\bar{x}$:

$$\begin{aligned} \bar{x} &= \frac{3.133\text{g} + 3.083\text{g} + \cdots + 2.520\text{g}}{30} \\ &= 2.804\text{g} \end{aligned}$$

3 The *sample standard deviation*, S:

$$\begin{aligned} S &= \sqrt{\frac{(3.133\text{g} - 2.804\text{g})^2 + (3.083 - 2.804\text{g})^2 + \cdots + (2.520 - 2.804\text{g})^2}{29}} \\ &= 0.3024\text{g} \end{aligned}$$

Year	Mass [g]
1989	2.516
1990	2.500
1991	2.500
1992	2.500
1993	2.503
1994	2.500
1995	2.497
1996	2.500
1997	2.494
1998	2.512
1999	2.521
2000	2.499
2001	2.523
2002	2.518
2003	2.520

Table 8.5: Mass of Pennies from 1989 to 2003.

4 The scale factor, k, adjusting for the small number of data points:

$$\begin{aligned} k &= 1 + \frac{20}{30^2} \\ &= 1.0222 \end{aligned}$$

Finally, we have the best estimate and uncertainty for the pennies in this full dataset:

$$\begin{aligned} \hat{\mu} &= \bar{x} \pm k \cdot S/\sqrt{N} \\ &= 2.804\text{g} \pm 1.0222 \cdot 0.3024\text{g}/\sqrt{30} \\ &= 2.804\text{g} \pm 0.0564\text{g} \end{aligned}$$

or, as a 99% credible range (3 times the uncertainty written above), we have,

$$\begin{aligned} 99\% \text{ CI for } \mu &= 2.804\text{g} \pm 3 \times 0.0564\text{g} \\ &= [2.634\text{g}, 2.973\text{g}] \end{aligned}$$

There are several things one should notice:

1 The scale factor, k, is less for 30 data points than it is for 15 data points. This is because the adjustment for small number of data points gets less relevant as we obtain more data. This is what we expect.

2 Despite there being twice as much data, our uncertainty *increased*. This is unusual, if our model is correct - more data should *sharpen*

the estimates. Although it is possible that adding more data increased the system variability somehow, it is more likely that some assumption of our model is incorrect. This becomes obvious when we look at the result graphically, shown in Figure 8.5.

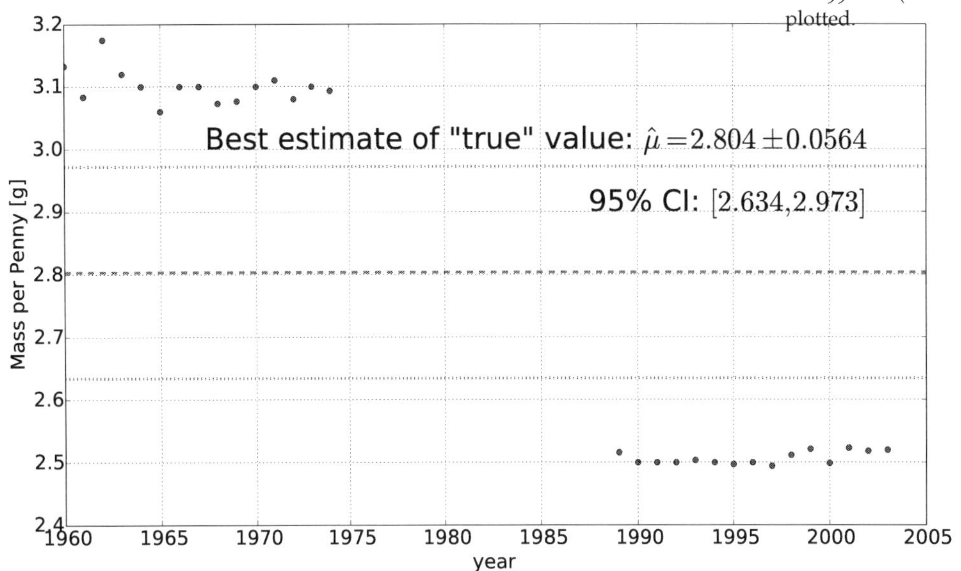

Figure 8.5: Mass of Pennies from 1960 to 2003, with best estimates and 99% CI (i.e. 3σ) uncertainty plotted.

This should highlight a few things:

1 Always look at your data graphically. What you might miss looking at a table of numbers, you'll catch with a picture.

2 Assume your model is wrong, and outline other possible models ahead of time and explore them. The most obvious improvement in this problem is to notice that we are dealing with two separate "true" values, possibly caused by a change in the manufacturing materials.

Example 8.14 *Mass of the Penny, Model 2 - Two True Values*

In this model, we assume there are two true values:

- μ_1 - before 1975
- μ_2 - after 1988

There are two roughly equivalent ways of telling whether there is a significant difference.

Overlapping Intervals The first is the easiest to do mathematically, and yields a nice picture: obtain the best estimates for μ_1 and μ_2, and see if their 99% credible intervals overlap. From this analysis (identical to the previous examples, however we leave the details of the calculation to the student), we get (see Figure 8.6):

- Best estimate for μ_1

$$\hat{\mu}_1 = 3.100 \pm 0.0078$$

  with 99% CI: [3.077,3.124].

- Best estimate for μ_2

$$\hat{\mu}_2 = 2.507 \pm 0.0029$$

  with 99% CI: [2.498,2.516]

where the 99% credible intervals (CI) clearly do not overlap, thus there is a statistically significant difference between them.

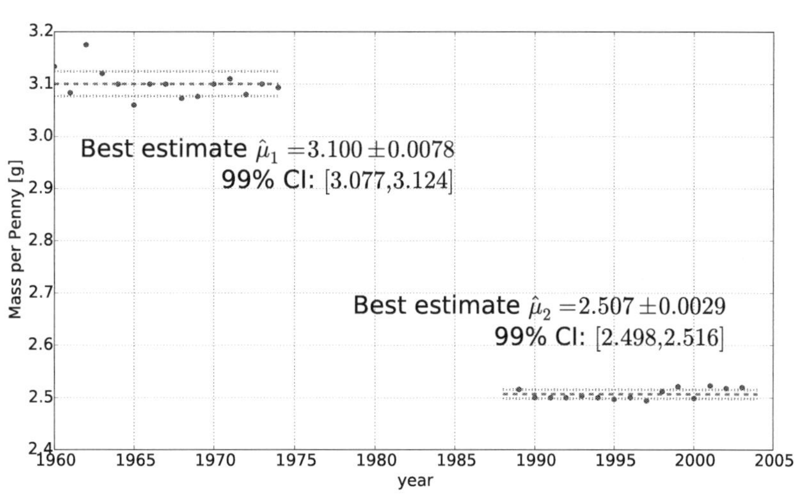

Figure 8.6: Mass of Pennies from 1960 to 2003, with best estimates for the two true values and their 99% CI (i.e. 3σ) uncertainty plotted. There is clearly no overlap in their credible intervals, thus there is a statistically significant difference between them.

Is the Difference Zero? The proper way is to estimate the quantity $\mu_1 - \mu_2$ and test to see if it is greater than zero, as shown in Section 7.2 on page 136. The estimate of this quantity, which we'll call $\delta_{12} = \mu_1 - \mu_2$ is related to the means and uncertainties of two data sets

$$\hat{\delta}_{12} = \bar{x}_1 - \bar{x}_2 \pm \sigma_{12}$$

$$\begin{aligned}
\sigma_{12} &= \sqrt{\sigma_1^2 + \sigma_2^2} \\
\sigma_1 &= k_1 S_1 / \sqrt{N_1} \text{ (uncertainty from data set 1)} \\
\sigma_2 &= k_2 S_2 / \sqrt{N_2} \text{ (uncertainty from data set 2)}
\end{aligned}$$

where the sample standard deviations, S_1 and S_2, and the scale factors, k_1 and k_2 were calculated earlier. This leads to, for this data set,

$$\hat{\delta}_{12} = 0.593\text{g} \pm 0.008\text{g}$$

with the 99% credible interval $[0.568\text{g}, 0.618\text{g}]$, the distribution shown in Figure 8.7. Again, the estimated quantities are clearly different statistically: the value of zero is *well* outside of the 99% credible interval for δ_{12}.

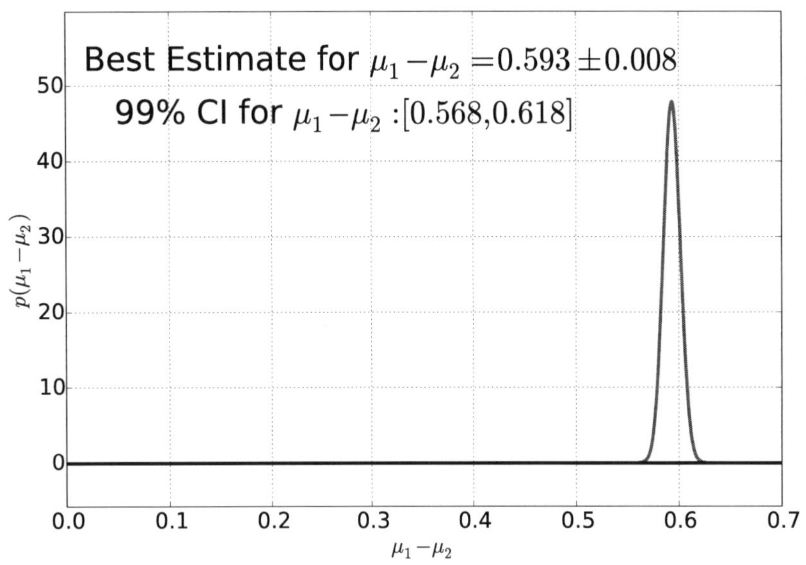

Figure 8.7: Difference in the estimated values of the pre- and post 1975 pennies, $\mu_1 - \mu_2$. The value zero is clearly outside of the 99% interval of the difference, thus there is a statistically significant difference between the two values μ_1 and μ_2.

8.5 *Computer Examples*

```
from sie import *
```

Iris Example

```
data=load_data('data/iris.csv')
```

```
x_sertosa=data[data['class']=='Iris-setosa']['petal length [cm]']
x_virginica=data[data['class']=='Iris-virginica']['petal length [cm]']
x_versicolor=data[data['class']=='Iris-versicolor']['petal length [cm]']
```

```
print x_sertosa[:10]  # print the first 10
```

```
0    1.4
1    1.4
2    1.3
3    1.5
4    1.4
5    1.7
6    1.4
7    1.5
8    1.4
9    1.5
Name: petal length [cm], dtype: float64
```

```
x=x_sertosa
mu=sample_mean(x)
N=len(x)
sigma=sample_deviation(x)/sqrt(N)
t_sertosa=tdist(N,mu,sigma)

print "total number of data points:",N
print "best estimate:",mu
print "uncertainty:",sigma
```

```
total number of data points: 50
best estimate: 1.464
uncertainty: 0.0245381834898
```

```
x=x_versicolor
mu=sample_mean(x)
N=len(x)
sigma=sample_deviation(x)/sqrt(N)
t_versicolor=tdist(N,mu,sigma)

print "total number of data points:",N
print "best estimate:",mu
print "uncertainty:",sigma
```

```
total number of data points: 50
best estimate: 4.26
uncertainty: 0.0664554477121
```

```
x=x_virginica
mu=sample_mean(x)
N=len(x)
sigma=sample_deviation(x)/sqrt(N)
t_virginica=tdist(N,mu,sigma)

print "total number of data points:",N
print "best estimate:",mu
print "uncertainty:",sigma
```

```
total number of data points: 50
best estimate: 5.552
uncertainty: 0.078049696361
```

```
distplot2([t_sertosa,t_versicolor,t_virginica],show_quartiles=False)
```

```
<matplotlib.figure.Figure at 0x1058d9690>
```

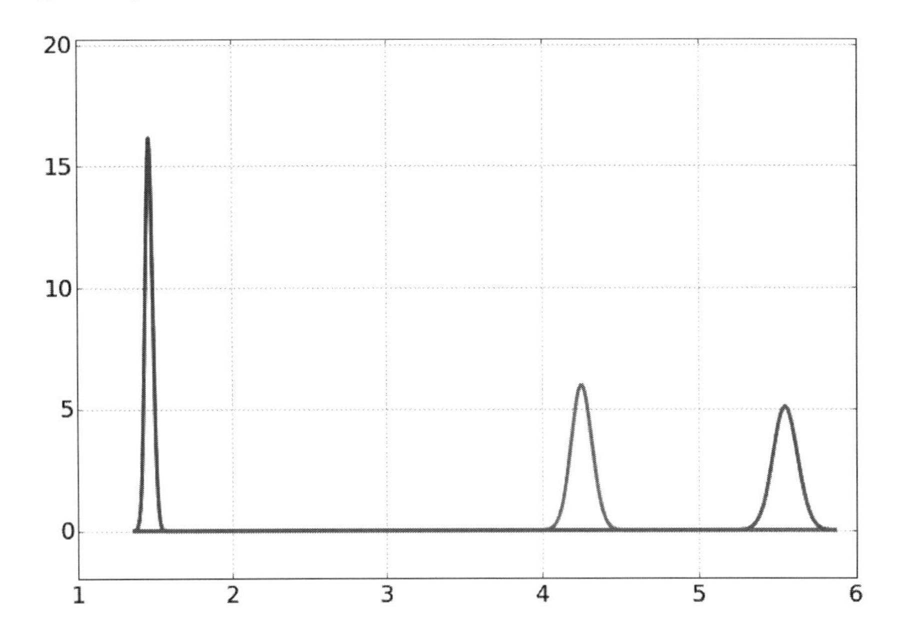


```
distplot(t_virginica)
```

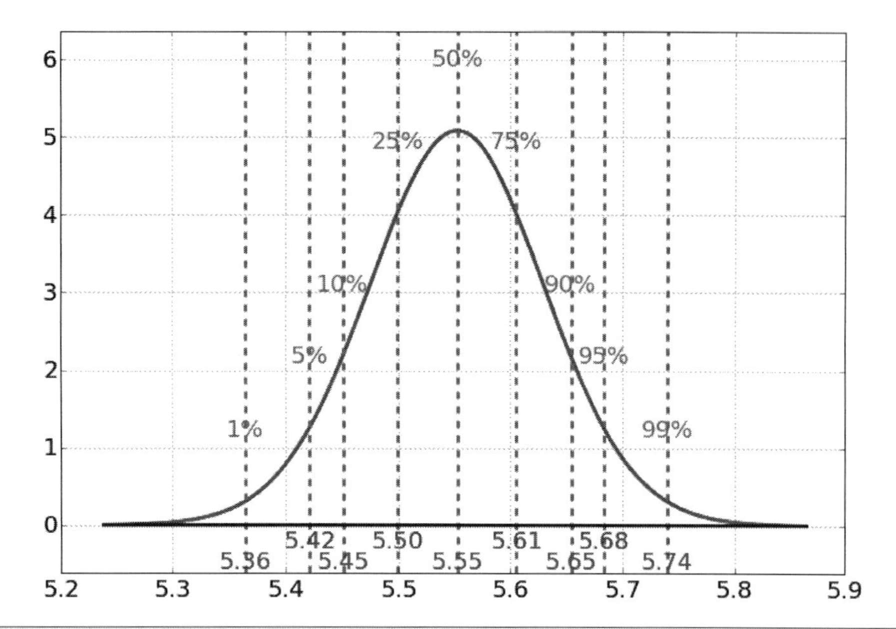


```
credible_interval(t_versicolor)
```

```
(4.1265203051077082, 4.2599999999999998, 4.3934796948922914)
```

```
credible_interval(t_virginica)
```

```
(5.3952325713636533, 5.5519999999999996, 5.7087674286363459)
```

Sunrise

```
dist=beta(h=365,N=365)
```

```
distplot(dist)
```

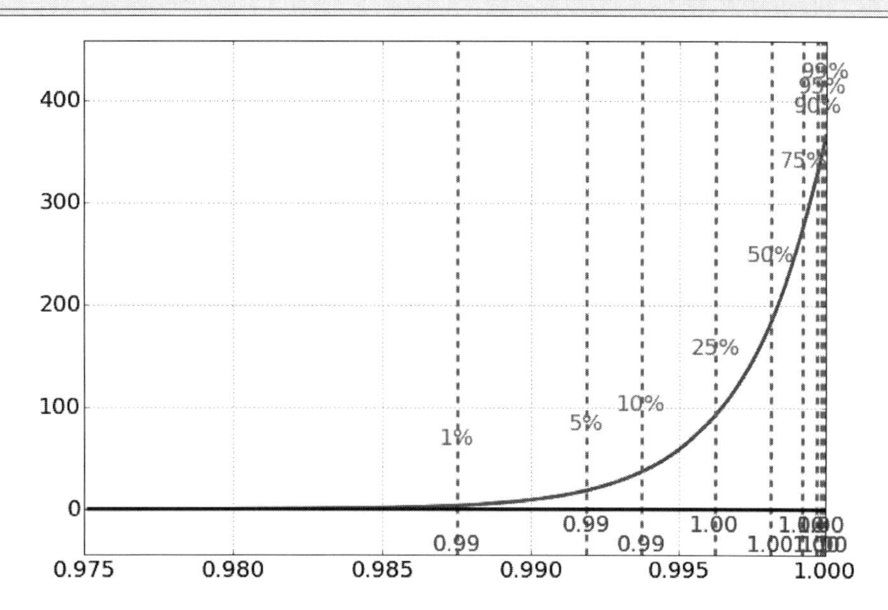

```
credible_interval(dist)
```

```
(0.98997171634278669, 0.99810794743679487, 0.99993082805373457)
```

Cancer Example

```
dist=beta(h=7,N=10)
```

```
distplot(dist,figsize=(8,5))
```

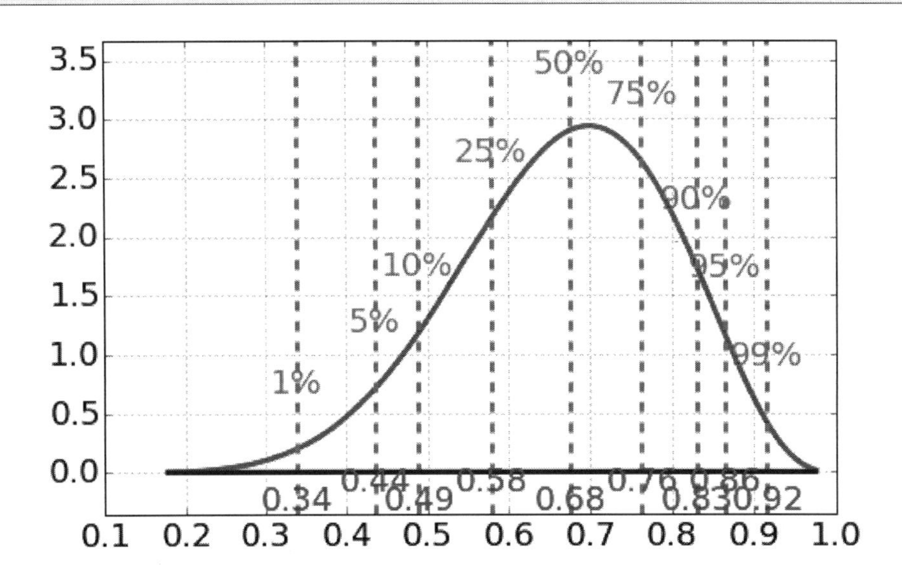


```
credible_interval(dist)
```

```
(0.39025744042757882, 0.67619553741481253, 0.89073655618090186)
```

Essentially no evidence of any effect over 50 percent.

Pennies

```
data1=load_data('data/pennies1.csv')
print data1
year,mass=data1['Year'],data1['Mass [g]']
```

```
   Year  Mass [g]
0  1960     3.133
1  1961     3.083
2  1962     3.175
3  1963     3.120
4  1964     3.100
5  1965     3.060
```

```
6   1966    3.100
7   1967    3.100
8   1968    3.073
9   1969    3.076
10  1970    3.100
11  1971    3.110
12  1972    3.080
13  1973    3.100
14  1974    3.093
```

```
plot(year,mass,'o')
xlabel('year')
ylabel('Mass per Penny [g]')
```

```
<matplotlib.text.Text at 0x1087c2d90>
```

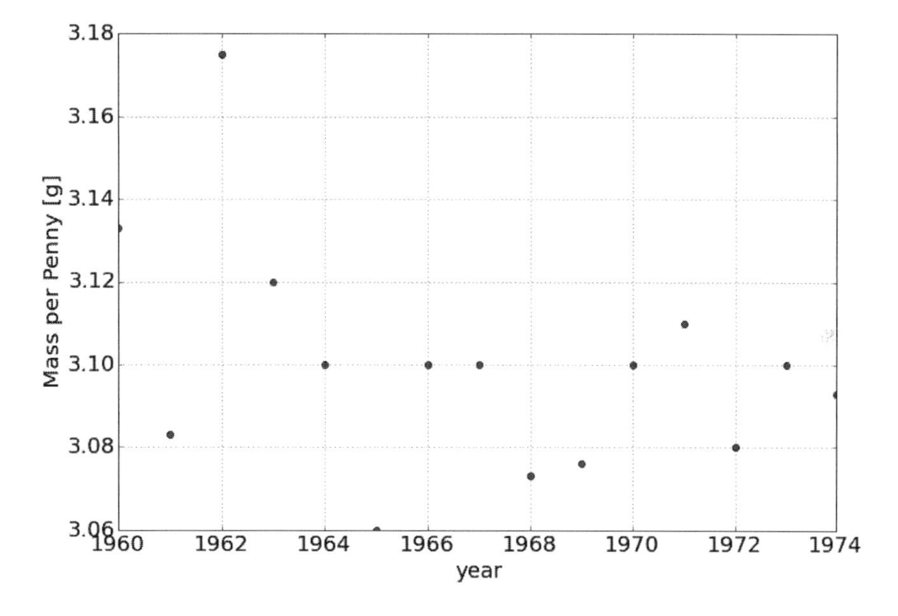


```
x=mass
mu=sample_mean(x)
N=len(x)
sigma=sample_deviation(x)/sqrt(N)
t_penny1=tdist(N,mu,sigma)

distplot(t_penny1,label='mass [g]')
```

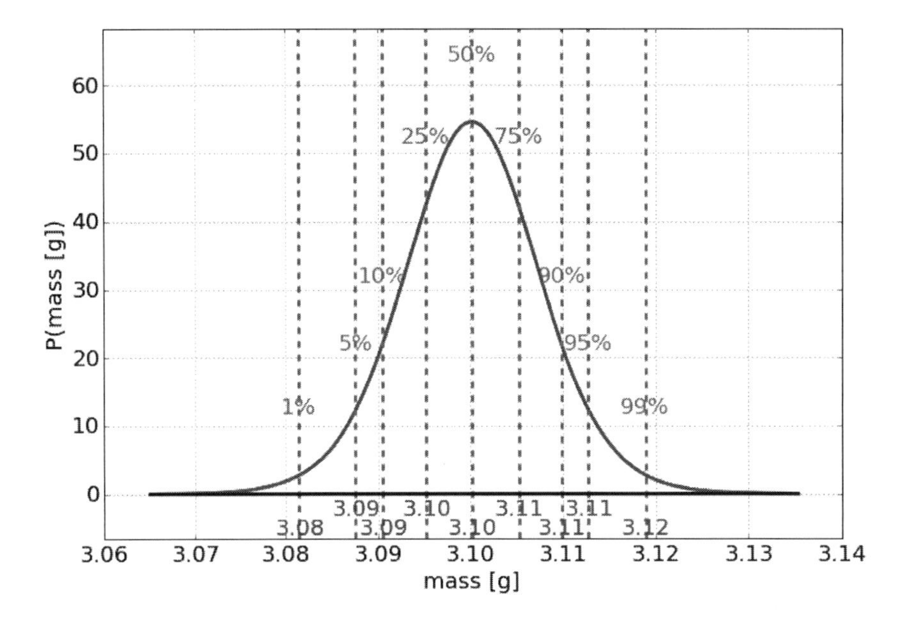


```
CI=credible_interval(t_penny1,percentage=99)
print CI
```

```
(3.0790129206702002, 3.1002000000000001, 3.1213870793298)
```

```
plot(year,mass,'o')
credible_interval_plot(t_penny1,percentage=99)
xlabel('year')
ylabel('Mass per Penny [g]')
```

```
<matplotlib.text.Text at 0x1087fcf10>
```

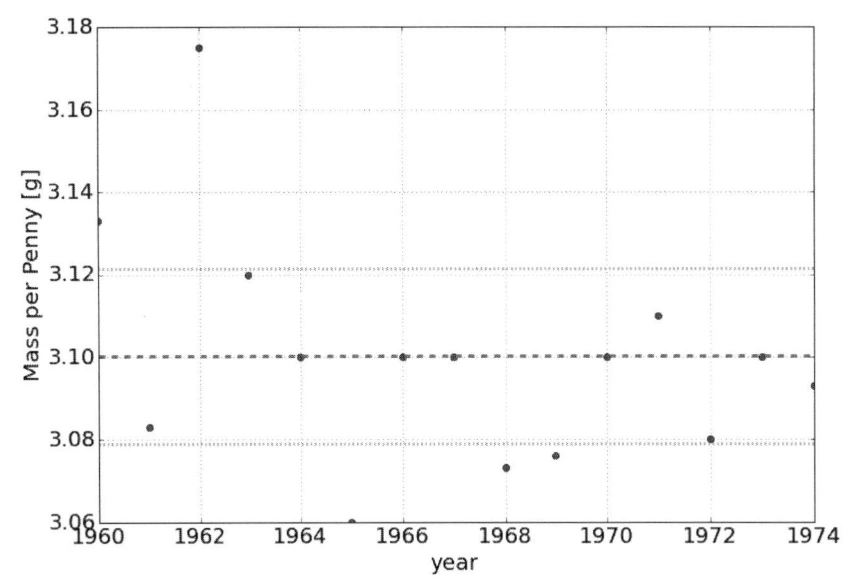

Do the 2 datasets

```
data2=load_data('data/pennies2.csv')
print data2
year1,mass1=year,mass
year2,mass2=data2['Year'],data2['Mass [g]']
```

```
    Year  Mass [g]
0   1989     2.516
1   1990     2.500
2   1991     2.500
3   1992     2.500
4   1993     2.503
5   1994     2.500
6   1995     2.497
7   1996     2.500
8   1997     2.494
9   1998     2.512
10  1999     2.521
11  2000     2.499
12  2001     2.523
13  2002     2.518
14  2003     2.520
```

```
x=mass1
mu=sample_mean(x)
N=len(x)
sigma=sample_deviation(x)/sqrt(N)
t_penny1=tdist(N,mu,sigma)

x=mass2
mu=sample_mean(x)
N=len(x)
sigma=sample_deviation(x)/sqrt(N)
t_penny2=tdist(N,mu,sigma)

distplot2([t_penny1,t_penny2],show_quartiles=False,label='mass [g]')
legend([r'$\mu_1$',r'$\mu_2$'])
```

```
<matplotlib.figure.Figure at 0x1087d3f10>

<matplotlib.legend.Legend at 0x1088198d0>
```

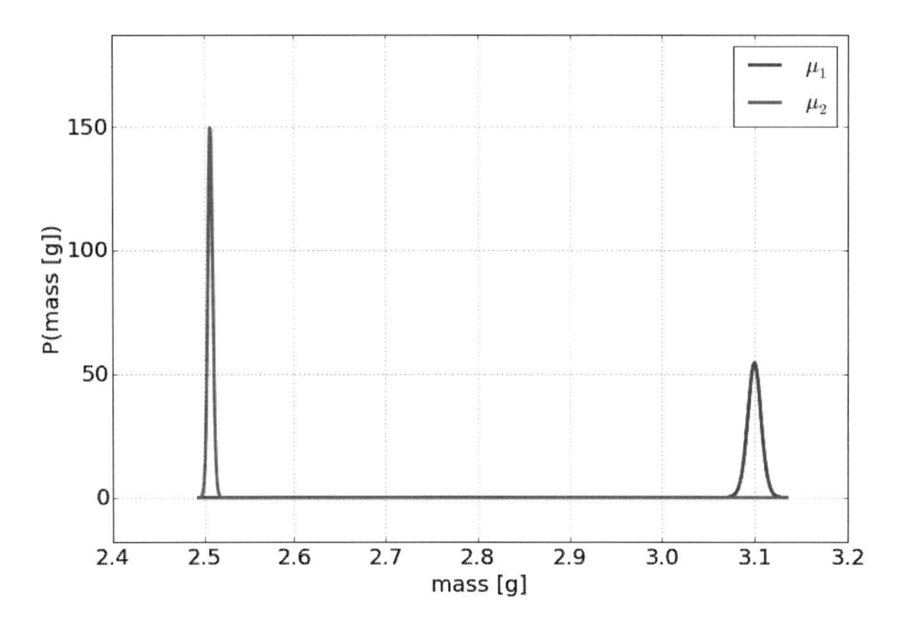


```
plot(year1,mass1,'o')
credible_interval_plot(t_penny1,percentage=99)
plot(year2,mass2,'ro')
credible_interval_plot(t_penny2,percentage=99,xlim=[1989,2005])
xlabel('year')
ylabel('Mass per Penny [g]')
```

```
<matplotlib.text.Text at 0x10907e310>
```

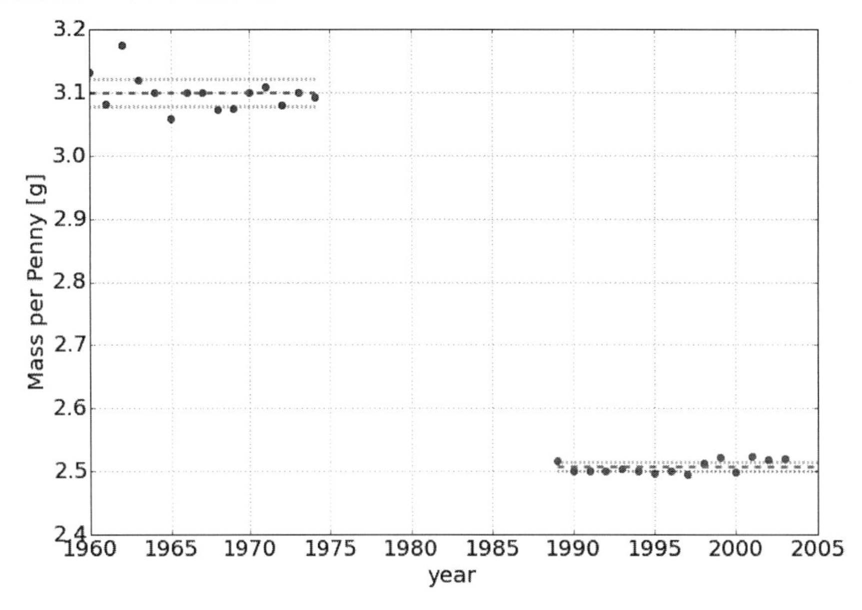

Distribution of the difference, normal approximation

```
N1=len(mass1)
N2=len(mass2)
```

```
mu1=sample_mean(mass1)
mu2=sample_mean(mass2)


sigma1=(1+20.0/N1**2)*sample_deviation(mass1)/sqrt(N1)
sigma2=(1+20.0/N2**2)*sample_deviation(mass2)/sqrt(N1)


delta_12=mu1-mu2
sigma_delta12=sqrt(sigma1**2+sigma2**2)

dist_delta=normal(delta_12,sigma_delta12)
distplot(dist_delta)
```

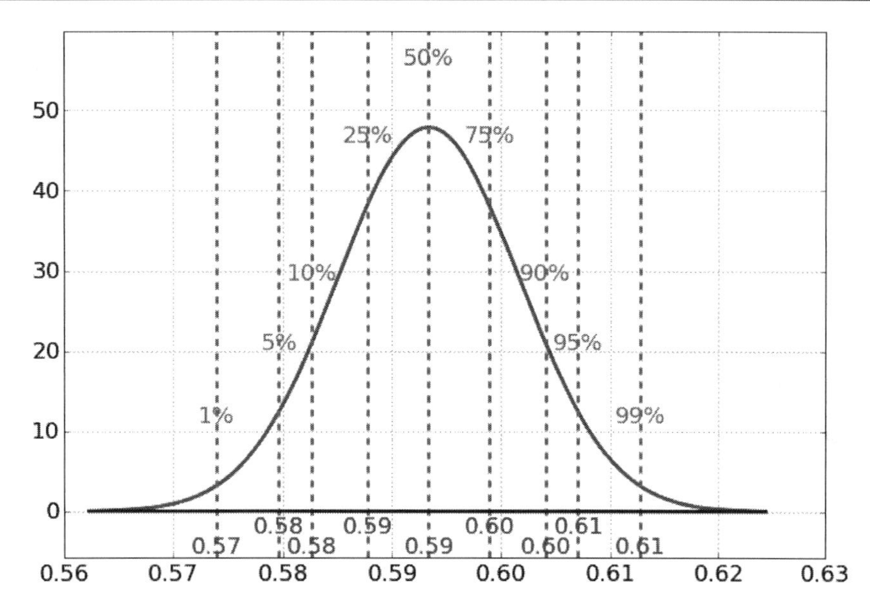

clearly larger than zero at well over the 99

Ball Bearing Sizes

```
data1=[1.18,1.42,0.69,0.88,1.62,1.09,1.53,1.02,1.19,1.32]
data2=[1.72,1.62,1.69,0.79,1.79,0.77,1.44,1.29,1.96,0.99]
N1=len(data1)
N2=len(data2)
```

```
mu1=sample_mean(data1)
mu2=sample_mean(data2)
print mu1,mu2
```

```
1.194 1.406
```

```
S1=sample_deviation(data1)
S2=sample_deviation(data2)
print S1,S2
```

```
0.289681817786 0.428309337849
```

```
sigma1=S1/sqrt(N1)
sigma2=S2/sqrt(N2)
print sigma1,sigma2
```

```
0.091605434094 0.135443305072
```

```
dist1=normal(mu1,sigma1)
dist2=normal(mu2,sigma2)
distplot2([dist1,dist2],show_quartiles=False,label='size [microns]')
legend([r'$\mu_1$',r'$\mu_2$'])
```

```
<matplotlib.figure.Figure at 0x105d61390>
<matplotlib.legend.Legend at 0x108ca1ad0>
```

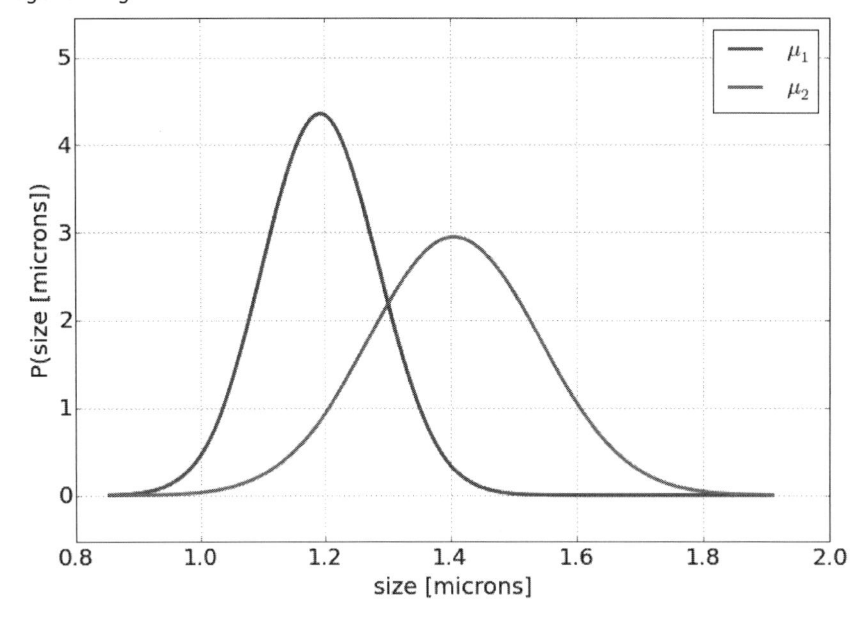

9 Concluding Thoughts

9.1 Where have we come?

We have tried in this book to present a particular picture of the world: everything is probability. We started with basic definitions and applications, and followed the consequences of the rules of probability to examine more complex problems. It is our hope that the reader sees that all of the analysis stems from a *single* perspective. In this way, one can approach *any* problem of inference in a unified way, applying the recipe we've used throughout:

1. Propose a model for the data you observe (which could be as simple as "there is an unknown true value for the observations")

2. Specify your prior knowledge of the parameters in the model, in the form of a prior probability (which is often as simple as "I don't know anything about the parameters, so all possible values are equally likely")

3. Specify how likely your data would be if your model were true, which is the likelihood part of Bayes' rule

4. Apply the rules of probability, namely Bayes' rule, to determine the posterior probability for the parameters in the model

5. Use the properties of probability functions to calculate answers to specific questions, for example "is it likely that this number is greater than zero?" or "are these two measurements different?"

Although I haven't covered all possible examples, and there are additions and clarifications still planned, this approach can be used for all new problems one faces. The only steps that can be daunting, at times, is the mathematical consequences and even there we have seen that the judicious use of approximations can go a long way.

9.2 *Where are we going?*

Topics I'd love to add, and will when I have the chance, include (in no particular order),

- Measurement in Science
- Linear Regression and Correlation
- Two-sample inferences
- Classification
- Model Building in Science
- Analysis of Social Science Data
- Inference for Deviation Parameters
- Experimental Design
- Computer simulations (e.g. MCMC)

Bibliography

Calif. couple wins two lotteries in one day, 2002. URL http://abcnews.go.com/US/story?id=90981.

Paul the octopus, July 2012. URL http://en.wikipedia.org/wiki/Psychic_octopus.

Alan Agresti and Brian Caffo. Simple and effective confidence intervals for proportions and differences of proportions result from adding two successes and two failures. *The American Statistician*, 54 (4):280–288, 2000.

Vincent Arel-Bundock. Rdatasets R datasets: An archive of datasets distributed with R, 2014. URL http://vincentarelbundock.github.io/Rdatasets/.

K. Bache and M. Lichman. UCI machine learning repository, 2013. URL http://archive.ics.uci.edu/ml.

D. Berry. *Statistics: A Bayesian Perspective.* Duxbury, 1996.

F. Galton. *Hereditary Genius: An Inquiry Into Its Laws and Consequences.* Macmillan and Company, limited, 1914. URL http://books.google.com/books?id=bJB9AAAAMAAJ.

A. Gelman, J. Hill, and Ebooks Corporation. *Data analysis using regression and multilevel/hierarchical models*, volume 625. Cambridge University Press Cambridge, UK:, 2007.

Andrew Gelman and Deborah Nolan. *Teaching Statistics: A Bag of Tricks*. Oxford University Press, USA, 2002. ISBN 0198572247. URL http://www.amazon.com/Teaching-Statistics-A-Bag-Tricks/dp/0198572247.

David J Hand, Fergus Daly, K McConway, D Lunn, and E Ostrowski. *A handbook of small data sets*, volume 1. CRC Press, 2011.

L Heaps. *Operation morning light*. Paddington, S.l, 1978. ISBN 0709203233.

E. T. Jaynes. *Probability Theory: The Logic of Science*. Cambridge University Press, Cambridge, 2003. Edited by G. Larry Bretthorst.

Lord Justice Kay. R vs Sally Clark, April 2003. URL http://www.bailii.org/ew/cases/EWCA/Crim/2003/1020.html.

D. V. Lindley and L. D. Phillips. Inference for a bernoulli process (a bayesian view). *The American Statistician*, 30(3):112–119, 1976.

Sharon McGrayne. *The Theory That Would Not Die: How Bayes' Rule Cracked the Enigma Code, Hunted Down Russian Submarines, and Emerged Triumphant from Two Centuries of Controversy*. Yale University Press, 2011. ISBN 0300169698.

L. Mlodinow. *The drunkard's walk: How randomness rules our lives*. Pantheon, 2008.

F. Mosteller. *Fifty challenging problems in probability with solutions*. Dover Pubns, 1965.

James Oberg. U.S. satellite shootdown: The inside story. *IEEE Spectrum*, 2008.

J. Randi. *Flim-flam!: psychics, ESP, unicorns, and other delusions*, volume 342. Prometheus Books Amherst, NY, 1982.

Carl Sagan. *Demon-Haunted World: Science as a Candle in the Dark*. Random House LLC, 1996.

S. Selvin. A problem in probability. *American Statistician*, 29(1):67, 1975.

J. Sullivan. People v. Collins , 68 cal.2d 319, 1968. URL http://scocal.stanford.edu/opinion/people-v-collins-22583.

A. Tversky and T. Gilovich. The cold facts about the" hot hand" in basketball. *Anthology of statistics in sports*, 16:169, 2005.

A. Tversky and D. Kahneman. Judgment under uncertainty: Heuristics and biases. *Science*, 185(4157):1124, 1974.

A. Tversky and D. Kahneman. Extensional versus intuitive reasoning: The conjunction fallacy in probability judgment. *Psychological review*, 90(4):293, 1983.

M. Vos Savant. Ask Marilyn [column]. *Parade Magazine*, page 16, 1990.

Wikipedia. Prosecutor's fallacy — Wikipedia, the free encyclopedia, 2014. URL http://en.wikipedia.org/wiki/Prosecutor's_fallacy.

Appendix A
Computational Analysis

The book is written with an accompanying software package, written in Python. As of this writing the recommended distribution for installing python is the Anaconda distribution, available here:

https://store.continuum.io/cshop/anaconda/

It is

- Free
- Easy to Use
- Easy to Extend
- Very Powerful

The accompanying software for the book can be obtained from the book website, http://web.bryant.edu/~bblais/statistical-inference-for-everyone-sie.html

Appendix B
Notation and Standards

B.1 Useful Greek Letters

Symbol	Name	Usage
α	Alpha	slope of a line
β	Beta	slope of a line, intercept
γ	Gamma	
Γ	Gamma	
δ	Delta	A small change in a variable
Δ	Delta	A change in a variable
ϵ	Epsilon	
ζ	Zeta	
η	Eta	
θ	Theta	The parameters in a binomial/-beta distribution
Θ	Theta	
κ	Kappa	
λ	Lambda	the mean in a poisson distribution
Λ	Lambda	
μ	Mu	the mean in a normal distribution (pronounced "mew")
ν	Nu	(pronounced "new")
ξ	Xi	
Ξ	Xi	
π	Pi	Represents the constant $3.1415\cdots$, the ratio of the circumference to the diameter of a circle
Π	Pi	A product of a series of numbers
ρ	Rho	
σ	Sigma	The standard width parameter of the normal distribution
Σ	Sigma	A sum of a series of numbers
τ	Tau	
ϕ	Phi	
Φ	Phi	
χ	Chi	A distribution related to the sum of normally distributed variables
ψ	Psi	
Ψ	Psi	
ω	Omega	
Ω	Omega	

B.2 Some Math Notation

Variables

A set of values, labeled with subscripts...

$$
\begin{aligned}
x_1 &= 1 \\
x_2 &= 5
\end{aligned}
$$

$$x_3 = -3$$
$$x_4 = 2$$
$$x_5 = 8$$

referred collectively as x_i.

Sums

$$x_1 + x_2 + x_3 + x_4 + x_5 = 1 + 5 + (-3) + 2 + 8 = 13$$

is equivalent to

$$\sum_{i=1}^{5} x_i = 1 + 5 + (-3) + 2 + 8 = 13$$

Products

$$x_1 \cdot x_2 \cdot x_3 \cdot x_4 \cdot x_5 = 1 \cdot 5 \cdot (-3) \cdot 2 \cdot 8 = -240$$

is equivalent to

$$\prod_{i=1}^{5} x_i = 1 \cdot 5 \cdot (-3) \cdot 2 \cdot 8 = -240$$

Sample Mean

The sample mean of a set of numbers is defined as...

$$\bar{x} \equiv \frac{x_1 + x_2 + \cdots x_N}{N}$$

In the example above

$$\bar{x} \equiv \frac{x_1 + x_2 + x_3 + x_4 + x_5}{5} = 2\frac{3}{5}$$

It can also be written

$$\bar{x} \equiv \frac{\sum_{i=1}^{N} x_i}{N}$$

or

$$\bar{x} \equiv \frac{\sum_i x_i}{N}$$

Sample Standard Deviation

$$s^2 \equiv \frac{1}{N-1}\sum_{i=1}^{N}(x-\bar{x})^2$$

$$s \equiv \sqrt{\frac{1}{N-1}\sum_{i=1}^{N}(x-\bar{x})^2}$$

Although the justification for the $N-1$ part is beyond this book, one easy way to remember it is that the sample distribution of a set of numbers is an estimate for the σ parameter of the normal distribution, representing the *spread* of the data. You can think of the $N-1$ part as a check to keep you from doing the crazy thing of estimating a spread with only 1 data point!

Estimates

Any specific estimate of a parameter, such as θ, is denoted with a hat, such as $\hat{\theta}$.

Factorials

Factorials are defined as

$$N! = 1 \cdot 2 \cdot 3 \cdots (N-1) \cdot N$$

for example

$$5! = 1 \cdot 2 \cdot 3 \cdot 4 \cdot 5 = 120$$

The *N-choose-k* notation is a shorthand for the factorials that arise in binomial and Beta distributions.

$$\binom{N}{k} \equiv \frac{N!}{k!(N-k)!}$$

B.3 *Qualitative labels to probability values*

Rough guide for the conversion of qualitative labels to probability values used throughout the book.

term	probability
virtually impossible	1/1,000,000
extremely unlikely	0.01 (i.e. 1/100)
very unlikely	0.05 (i.e. 1/20)
unlikely	0.2 (i.e. 1/5)
slightly unlikely	0.4 (i.e. 2/5)
even odds	0.5 (i.e. 50-50)
slightly likely	0.6 (i.e. 3/5)
likely	0.8 (i.e. 4/5)
very likely	0.95 (i.e. 19/20)
extremely likely	0.99 (i.e. 99/100)
virtually certain	999,999/1,000,000

Appendix C
Common Distributions and Their Properties

This chapter is really a reference for the standard distributions encountered in statistical inference. Although you are encouraged to read this chapter through, it can also be read out-of-order to look at a specific distribution.

C.1 *Discrete and Continuous*

Some distributions apply to a discrete (i.e. countable) number of possibilities while others apply to continuous values. In the case of discrete variables, the probability is given by the actual value of the distribution, so it makes sense to speak of the probability of an individual label, $P(\text{coin1})$. In the case of continuous variables, the probability is given by the area under the distribution, so it makes sense only to speak of the probability if a range of labels, $P(0.2 < \theta < 0.3)$.

C.2 *Uniform*

Discrete

Discrete uniform distribution The discrete uniform distribution is defined to be a *constant* value for all possibilities. Mathematically this is written

$$p(x_i) = \frac{1}{N}$$

where N is the total number of possibilities, labeled x_1 to x_N. The picture of the distribution is shown in Figure C.1

Discrete uniform distribution The discrete uniform distribution is defined to be a *constant* value for all possibilities. Mathematically this is written

$$p(x_i) = \frac{1}{N}$$

where N is the total number of possibilities, labeled x_1 to x_N.

Continuous

Continuous uniform distribution The continuous uniform distribution is defined to be a *constant* between a minimum and maximum

Continuous uniform distribution The continuous uniform distribution is defined to be a *constant* between a minimum and maximum value, and zero everywhere else. Mathematically this is written

$$p(x) = \frac{1}{\max - \min} \text{ for } \min < x < \max$$

.

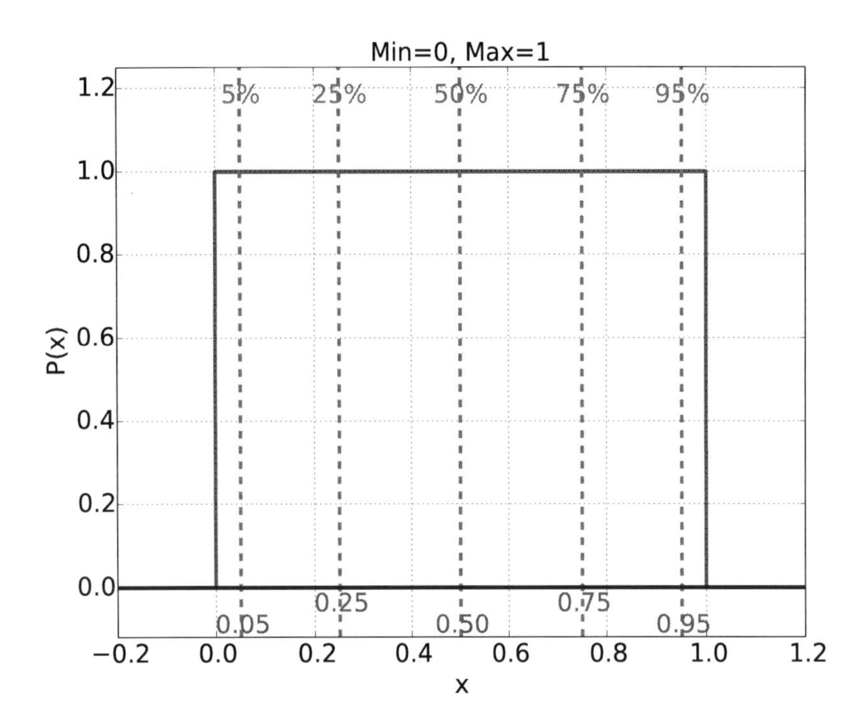

Figure C.1: Discrete uniform distribution for values 1 to 6. The value for each is $p(x_i) = 1/6$.

value, and zero everywhere else. Mathematically this is written

$$p(x) = \frac{1}{\max - \min} \text{ for } \min < x < \max$$

. The picture of the distribution is shown in Figure C.2.

EXAMPLE C.1 *You call a plumber, and they say that they can come anytime in the next 4 hours. The probability of them arriving at any particular time can be represented with a uniform distribution. What is the probability that they arrive in the first 20 minutes of the second hour?*

In order to ask questions about total probability from a *continuous* distribution you take the *area under the curve* between the relevant values. In this case it'd be the area under the curve from the time $t =$ 2hr and $t =$ 2hr + 20minutes = 2.333hr, as shown in Figure C.3. The area under the curve is just the area of the shaded region between times $t =$ 2hr and $t =$ 2.333hr, or just the area of a rectangle - $A =$ base $\times$ height. The *base* of the rectangle is the length of time, or

$$base = 0.333\text{hr}$$

The height of the rectangle is given by the constant value of the uniform distribution, or

$$height = \frac{1}{\max - \min} = \frac{1}{4\text{hr} - 0\text{hr}} = 0.25\frac{1}{\text{hr}}$$

The reason for the particular constant value for the uniform distribution, $1/(\max - \min)$, is simply that the area of the entire rectangle must be 1, which means that there is a 100% chance of the values falling between the minimum and maximum values.

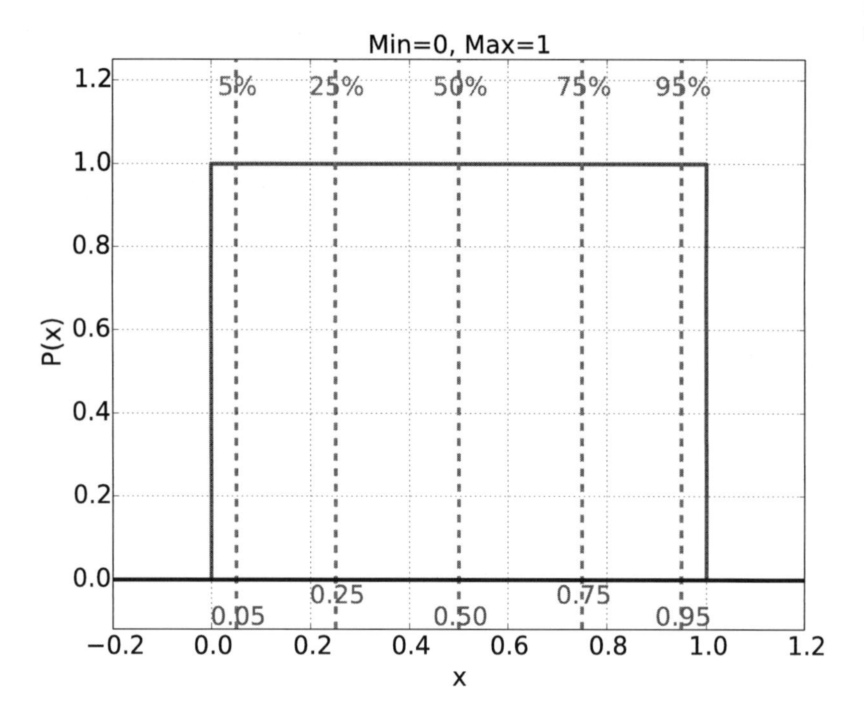

Figure C.2: Continuous uniform distribution between values 0 and 1

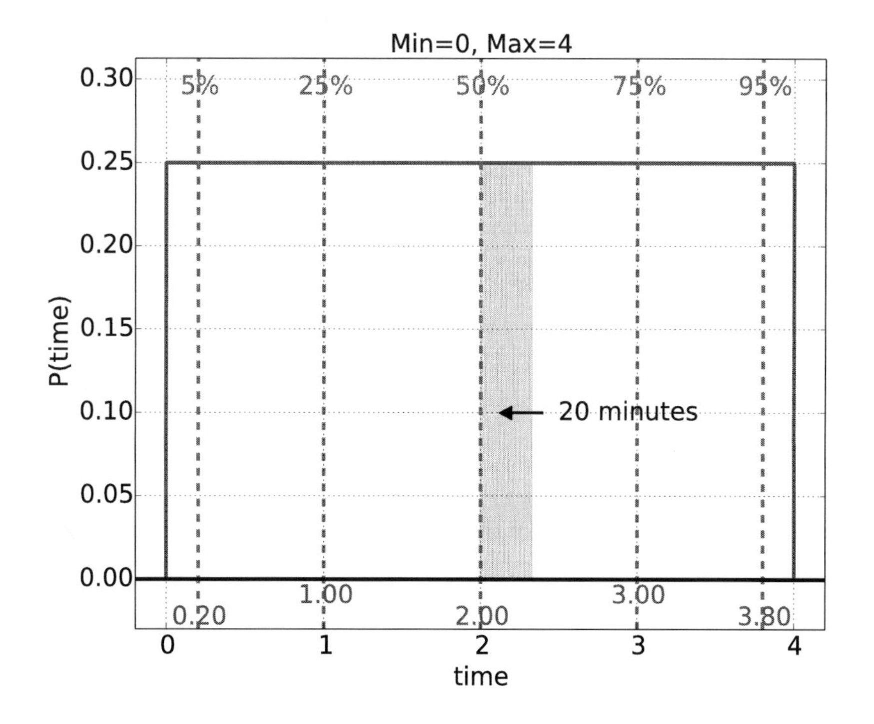

Figure C.3: Continuous uniform distribution for the plumber example (Example C.1).

So the total probability of the plumber coming in the first 20 minutes of the second hour is

$$P(2 < t < 2.25) = (0.333\text{hr}) \times \left(0.25\frac{1}{\text{hr}}\right) = 0.0833$$

C.3 Binomial

Binomial distribution The discrete binomial distribution is defined to be the probability of achieving h successes in a given N events where each event has a given θ probability of success.

$$P(h|N,\theta) = \left(\begin{array}{c} h \\ N \end{array}\right) \theta^h (1-\theta)^{N-h}$$

Binomial distribution The discrete binomial distribution is defined to be the probability of achieving h successes in a given N events where each event has a given θ probability of success.

$$P(h|N,\theta) = \left(\begin{array}{c} h \\ N \end{array}\right) \theta^h (1-\theta)^{N-h}$$

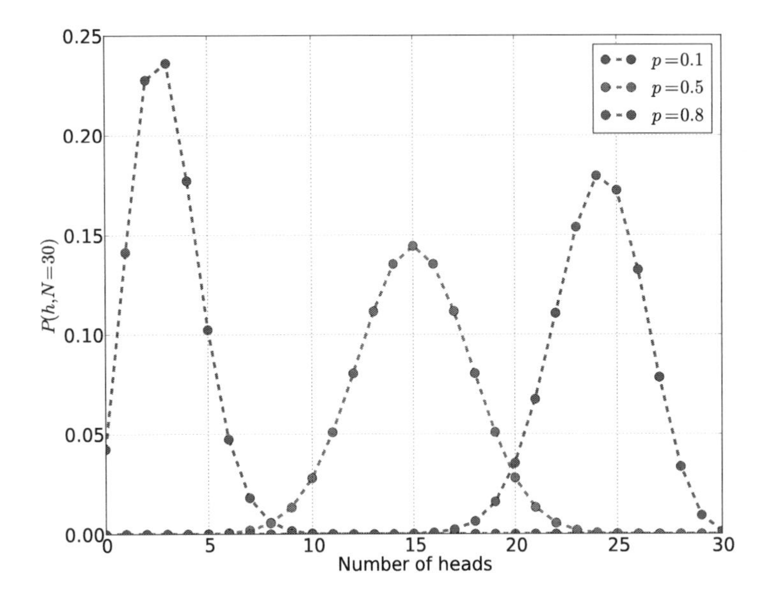

Figure C.4: Probability of getting h heads in 30 flips given a possible unfair coin. One coin has $p = 0.1$, where the maximum is for 3 heads (or 1/10 of the 30 flips), but 2 heads is nearly as likely. Another has $p = 0.5$, and is the fair coin considered earlier with a maximum at 15 heads (or 1/2 of the 30 flips). Finally, another coin shown as $p = 0.8$ where 24 heads (or 8/10 of the 30 flips) is maximum.

Although it may look like a Beta, the binomial distribution is used to find the best estimate for the number of successes, h, given the number of events, N, and the probability of the success of a single event, θ.

C.4 Beta

Beta distribution The continuous Beta distribution is the posterior probability distribution for the parameter θ, where one has observed

Beta distribution The continuous Beta distribution is the posterior probability distribution for the parameter θ, where one has observed h successes in a given N events, and each event is assumed to have a θ probability of success.

$$P(\theta|h,N) = (N+1)\cdot\left(\begin{array}{c} N \\ h \end{array}\right) \theta^h (1-\theta)^{N-h}$$

h successes in a given N events, and each event is assumed to have a θ probability of success.

$$P(\theta|h,N) = (N+1) \cdot \binom{N}{h} \theta^h (1-\theta)^{N-h}$$

Although it may look like a binomial, the Beta distribution is used to find the best estimate for the parameter θ where the number of successes and events, h and N are given.

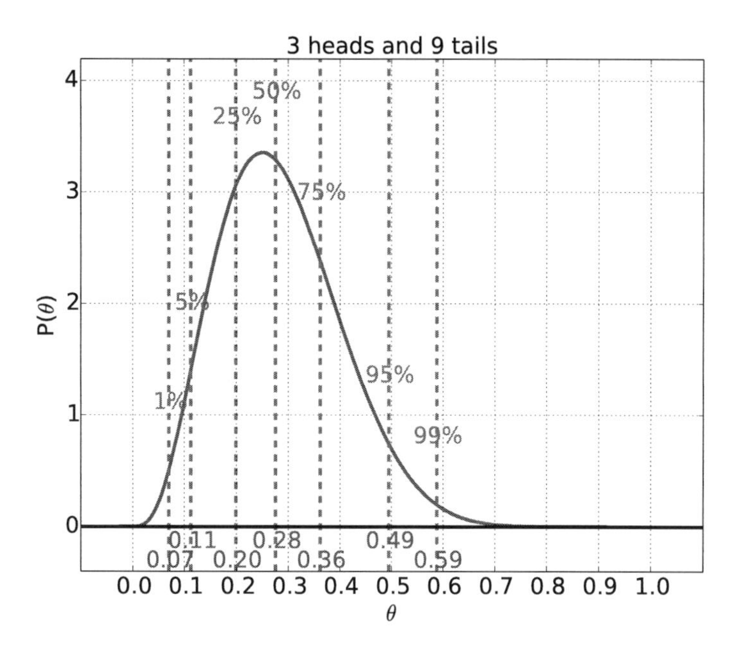

Figure C.5: Posterior probability distribution for the θ values of the bent coin - the probability that the coin will land heads. The distribution is shown for data 3 heads and 9 tails. The various quartiles are shown in the plot.

C.5 *Normal (Gaussian)*

Normal distribution The Normal distribution is the most common distribution found in all of statistical inference. It is the best prior distribution to use, when all you know is that your data has a constant true value and some constant variation around that true value. It is the posterior probability distribution for the unknown true value given N samples and the known deviation, σ. It is also the approximate form for nearly every distribution when you have many samples. The mathematical form for the normal, or Gaussian, is

$$\text{Normal}(\mu,\sigma) = \frac{1}{\sqrt{2\pi\sigma^2}} e^{-(x-\mu)^2/2\sigma^2}$$

Normal distribution The Normal distribution is the most common distribution found in all of statistical inference. It is the best prior distribution to use, when all you know is that your data has a constant true value and some constant variation around that true value. It is the posterior probability distribution for the unknown true value given N samples and the known deviation, σ. It is also the approximate form for nearly every distribution when you have many samples. The mathematical form for the normal, or Gaussian, is

$$\text{Normal}(\mu,\sigma) = \frac{1}{\sqrt{2\pi\sigma^2}} e^{-(x-\mu)^2/2\sigma^2}$$

Figure C.6: The normal distribution.

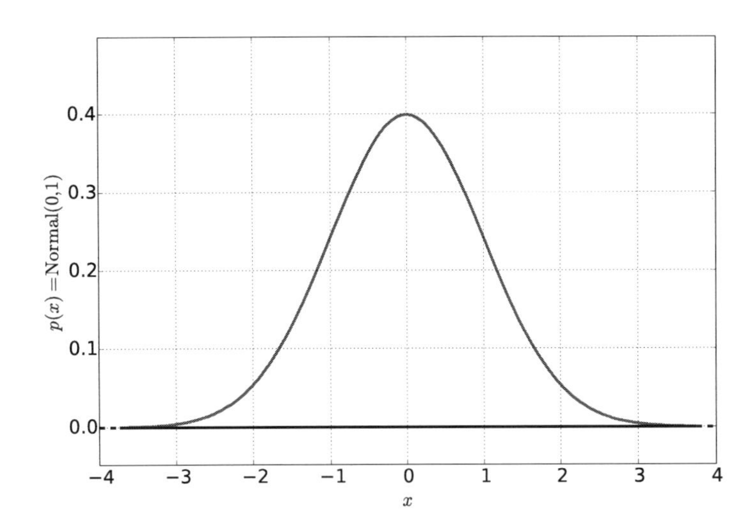

Three useful properties of σ for the normal distribution are the following:

1 the normal distribution value at the maximum (i.e. at $x = \mu$) is around 2.7 times larger than the value one-σ away from the maximum (at $x = \mu - \sigma$ and $x = \mu + \sigma$)

2 the total probability between these two points is 65%.

3 95% of the distribution lies between $\mu - 2\sigma$ and $\mu + 2\sigma$ (see Figure 7.3)

Appendix D Tables

D.1 Credible Intervals for Standard Normal Distribution

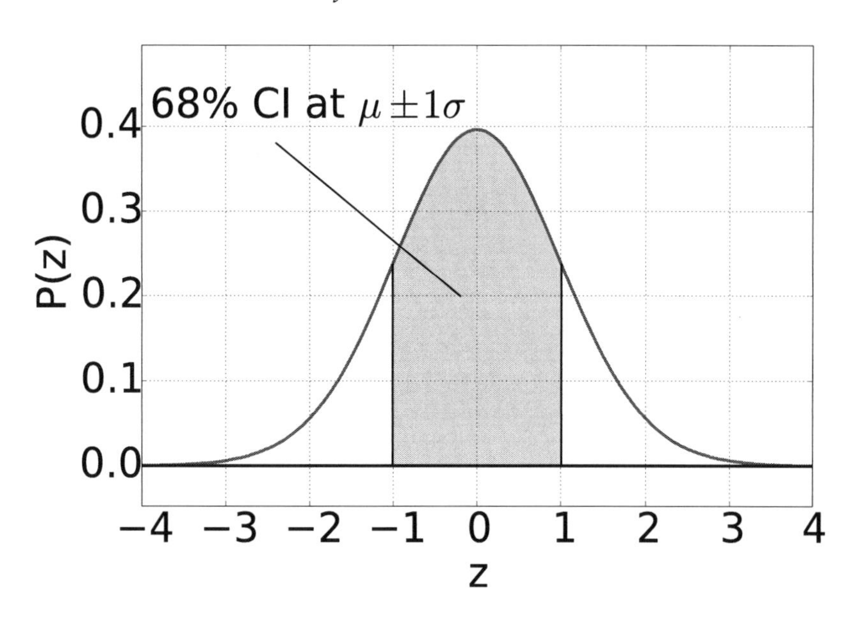

Credible Interval	$\pm$z	Approximately
50.0%	0.6745σ	
68.0%	0.9945σ	1σ
90.0%	1.6449σ	
95.0%	1.9600σ	2σ
99.0%	2.5758σ	
99.8%	3.0902σ	3σ
99.995%	4.0556σ	4σ

EXAMPLE D.1 *Usage of the Credible Interval Table for the Normal Distribution*

Given a set of 10 samples with sample mean $\bar{x} = 5.2$ and known deviation $\sigma = 0.3$, the best estimate for the mean parameter μ, representing the true value of the data, is the sample mean, $\hat{\mu} = 5.2$ with uncertainty $\sigma/\sqrt{N}$ or $0.3/\sqrt{10} = 0.095$. Some of the credible intervals for this estimate then are the following

- 68% - $[5.2 - 0.9945 \cdot 0.095, 5.2 + 0.9945 \cdot 0.095] = [5.11, 5.29]$
- 95% - $[5.2 - 1.9600 \cdot 0.095, 5.2 + 1.9600 \cdot 0.095] = [5.01, 5.39]$
- 99.8% - $[5.2 - 3.0902 \cdot 0.095, 5.2 + 3.0902 \cdot 0.095] = [4.91, 5.49]$

or approximately

- 68% - $[5.2 - 1 \cdot 0.095, 5.2 + 1 \cdot 0.095] = [5.11, 5.29]$
- 95% - $[5.2 - 2 \cdot 0.095, 5.2 + 2 \cdot 0.095] = [5.01, 5.39]$
- 99.8% - $[5.2 - 3 \cdot 0.095, 5.2 + 3 \cdot 0.095] = [4.91, 5.49]$

D.2 *Credible Intervals for Student's t Distribution*

	Degrees of Freedom							
Credible Interval	1	2	3	4	5	6	7	8
50.0%	1.000σ	0.816σ	0.765σ	0.741σ	0.727σ	0.718σ	0.711σ	0.706σ
68.0%	1.819σ	1.312σ	1.189σ	1.134σ	1.104σ	1.084σ	1.070σ	1.060σ
90.0%	6.314σ	2.920σ	2.353σ	2.132σ	2.015σ	1.943σ	1.895σ	1.860σ
95.0%	12.706σ	4.303σ	3.182σ	2.776σ	2.571σ	2.447σ	2.365σ	2.306σ
99.0%	63.657σ	9.925σ	5.841σ	4.604σ	4.032σ	3.707σ	3.499σ	3.355σ
99.8%	318.309σ	22.327σ	10.215σ	7.173σ	5.893σ	5.208σ	4.785σ	4.501σ
99.995%	12732.395σ	141.416σ	35.298σ	18.522σ	12.893σ	10.261σ	8.783σ	7.851σ

	Degrees of Freedom							
Credible Interval	9	10	11	12	13	14	15	16
50.0%	0.703σ	0.700σ	0.697σ	0.695σ	0.694σ	0.692σ	0.691σ	0.690σ
68.0%	1.053σ	1.046σ	1.041σ	1.037σ	1.034σ	1.031σ	1.029σ	1.026σ
90.0%	1.833σ	1.812σ	1.796σ	1.782σ	1.771σ	1.761σ	1.753σ	1.746σ
95.0%	2.262σ	2.228σ	2.201σ	2.179σ	2.160σ	2.145σ	2.131σ	2.120σ
99.0%	3.250σ	3.169σ	3.106σ	3.055σ	3.012σ	2.977σ	2.947σ	2.921σ
99.8%	4.297σ	4.144σ	4.025σ	3.930σ	3.852σ	3.787σ	3.733σ	3.686σ
99.995%	7.215σ	6.757σ	6.412σ	6.143σ	5.928σ	5.753σ	5.607σ	5.484σ

Credible Interval	Degrees of Freedom 17	18	19	20	21	22	23	24
50.0%	0.689σ	0.688σ	0.688σ	0.687σ	0.686σ	0.686σ	0.685σ	0.685σ
68.0%	1.024σ	1.023σ	1.021σ	1.020σ	1.019σ	1.017σ	1.016σ	1.015σ
90.0%	1.740σ	1.734σ	1.729σ	1.725σ	1.721σ	1.717σ	1.714σ	1.711σ
95.0%	2.110σ	2.101σ	2.093σ	2.086σ	2.080σ	2.074σ	2.069σ	2.064σ
99.0%	2.898σ	2.878σ	2.861σ	2.845σ	2.831σ	2.819σ	2.807σ	2.797σ
99.8%	3.646σ	3.610σ	3.579σ	3.552σ	3.527σ	3.505σ	3.485σ	3.467σ
99.995%	5.379σ	5.288σ	5.209σ	5.139σ	5.077σ	5.022σ	4.972σ	4.927σ

Credible Interval	Degrees of Freedom 25	26	27	28	29	30	31	32
50.0%	0.684σ	0.684σ	0.684σ	0.683σ	0.683σ	0.683σ	0.682σ	0.682σ
68.0%	1.015σ	1.014σ	1.013σ	1.012σ	1.012σ	1.011σ	1.011σ	1.010σ
90.0%	1.708σ	1.706σ	1.703σ	1.701σ	1.699σ	1.697σ	1.696σ	1.694σ
95.0%	2.060σ	2.056σ	2.052σ	2.048σ	2.045σ	2.042σ	2.040σ	2.037σ
99.0%	2.787σ	2.779σ	2.771σ	2.763σ	2.756σ	2.750σ	2.744σ	2.738σ
99.8%	3.450σ	3.435σ	3.421σ	3.408σ	3.396σ	3.385σ	3.375σ	3.365σ
99.995%	4.887σ	4.849σ	4.816σ	4.784σ	4.756σ	4.729σ	4.705σ	4.682σ

Credible Interval	Degrees of Freedom 33	34	35	36	37	38	39	40
50.0%	0.682σ	0.682σ	0.682σ	0.681σ	0.681σ	0.681σ	0.681σ	0.681σ
68.0%	1.010σ	1.009σ	1.009σ	1.008σ	1.008σ	1.008σ	1.007σ	1.007σ
90.0%	1.692σ	1.691σ	1.690σ	1.688σ	1.687σ	1.686σ	1.685σ	1.684σ
95.0%	2.035σ	2.032σ	2.030σ	2.028σ	2.026σ	2.024σ	2.023σ	2.021σ
99.0%	2.733σ	2.728σ	2.724σ	2.719σ	2.715σ	2.712σ	2.708σ	2.704σ
99.8%	3.356σ	3.348σ	3.340σ	3.333σ	3.326σ	3.319σ	3.313σ	3.307σ
99.995%	4.660σ	4.640σ	4.622σ	4.604σ	4.588σ	4.572σ	4.558σ	4.544σ

EXAMPLE D.2 *Usage of the Credible Interval Table for the Student's t Distribution*

Given a set of 10 samples (9 degrees of freedom) with sample mean $\bar{x} = 5.2$ and sample deviation $s = 0.3$, the best estimate for the mean parameter μ, representing the true value of the data, is the sample mean, $\hat{\mu} = 5.2$ with uncertainty $s/\sqrt{N}$ or $0.3/\sqrt{10} = 0.095$. Some of the credible intervals for this estimate then are the following

- 68% - $[5.2 - 1.053 \cdot 0.095, 5.2 + 1.053 \cdot 0.095] = [5.09, 5.3]$
- 95% - $[5.2 - 2.262 \cdot 0.095, 5.2 + 2.262 \cdot 0.095] = [4.99, 5.41]$
- 99.8% - $[5.2 - 4.297 \cdot 0.095, 5.2 + 4.297 \cdot 0.095] = [4.79, 5.61]$

D.3 *Cumulative Standard Normal Distribution*

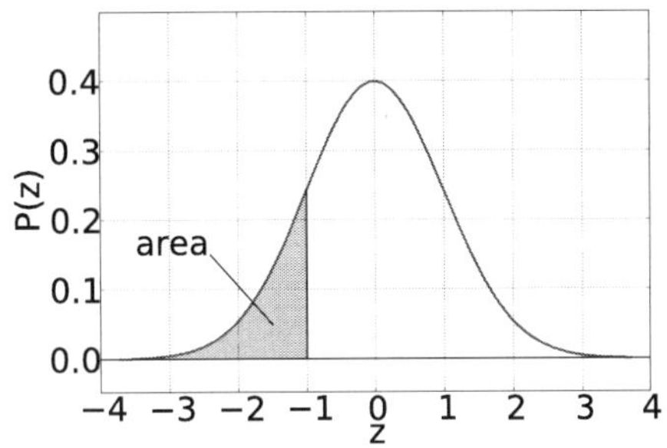

z	Area on Left	z	Area on Left	z	Area on Left	z	Area on Left	z	Area on Left
-3.70	0.0001	-3.40	0.0003	-3.10	0.0010	-2.80	0.0026	-2.50	0.0062
-3.69	0.0001	-3.39	0.0003	-3.09	0.0010	-2.79	0.0026	-2.49	0.0064
-3.68	0.0001	-3.38	0.0004	-3.08	0.0010	-2.78	0.0027	-2.48	0.0066
-3.67	0.0001	-3.37	0.0004	-3.07	0.0011	-2.77	0.0028	-2.47	0.0068
-3.66	0.0001	-3.36	0.0004	-3.06	0.0011	-2.76	0.0029	-2.46	0.0069
-3.65	0.0001	-3.35	0.0004	-3.05	0.0011	-2.75	0.0030	-2.45	0.0071
-3.64	0.0001	-3.34	0.0004	-3.04	0.0012	-2.74	0.0031	-2.44	0.0073
-3.63	0.0001	-3.33	0.0004	-3.03	0.0012	-2.73	0.0032	-2.43	0.0075
-3.62	0.0001	-3.32	0.0005	-3.02	0.0013	-2.72	0.0033	-2.42	0.0078
-3.61	0.0002	-3.31	0.0005	-3.01	0.0013	-2.71	0.0034	-2.41	0.0080
-3.60	0.0002	-3.30	0.0005	-3.00	0.0013	-2.70	0.0035	-2.40	0.0082
-3.59	0.0002	-3.29	0.0005	-2.99	0.0014	-2.69	0.0036	-2.39	0.0084
-3.58	0.0002	-3.28	0.0005	-2.98	0.0014	-2.68	0.0037	-2.38	0.0087
-3.57	0.0002	-3.27	0.0005	-2.97	0.0015	-2.67	0.0038	-2.37	0.0089
-3.56	0.0002	-3.26	0.0006	-2.96	0.0015	-2.66	0.0039	-2.36	0.0091
-3.55	0.0002	-3.25	0.0006	-2.95	0.0016	-2.65	0.0040	-2.35	0.0094
-3.54	0.0002	-3.24	0.0006	-2.94	0.0016	-2.64	0.0041	-2.34	0.0096
-3.53	0.0002	-3.23	0.0006	-2.93	0.0017	-2.63	0.0043	-2.33	0.0099
-3.52	0.0002	-3.22	0.0006	-2.92	0.0018	-2.62	0.0044	-2.32	0.0102
-3.51	0.0002	-3.21	0.0007	-2.91	0.0018	-2.61	0.0045	-2.31	0.0104
-3.50	0.0002	-3.20	0.0007	-2.90	0.0019	-2.60	0.0047	-2.30	0.0107
-3.49	0.0002	-3.19	0.0007	-2.89	0.0019	-2.59	0.0048	-2.29	0.0110
-3.48	0.0003	-3.18	0.0007	-2.88	0.0020	-2.58	0.0049	-2.28	0.0113
-3.47	0.0003	-3.17	0.0008	-2.87	0.0021	-2.57	0.0051	-2.27	0.0116
-3.46	0.0003	-3.16	0.0008	-2.86	0.0021	-2.56	0.0052	-2.26	0.0119
-3.45	0.0003	-3.15	0.0008	-2.85	0.0022	-2.55	0.0054	-2.25	0.0122
-3.44	0.0003	-3.14	0.0008	-2.84	0.0023	-2.54	0.0055	-2.24	0.0125
-3.43	0.0003	-3.13	0.0009	-2.83	0.0023	-2.53	0.0057	-2.23	0.0129
-3.42	0.0003	-3.12	0.0009	-2.82	0.0024	-2.52	0.0059	-2.22	0.0132
-3.41	0.0003	-3.11	0.0009	-2.81	0.0025	-2.51	0.0060	-2.21	0.0136
-3.40	0.0003	-3.10	0.0010	-2.80	0.0026	-2.50	0.0062	-2.20	0.0139

Cumulative Normal Distribution (cont.)

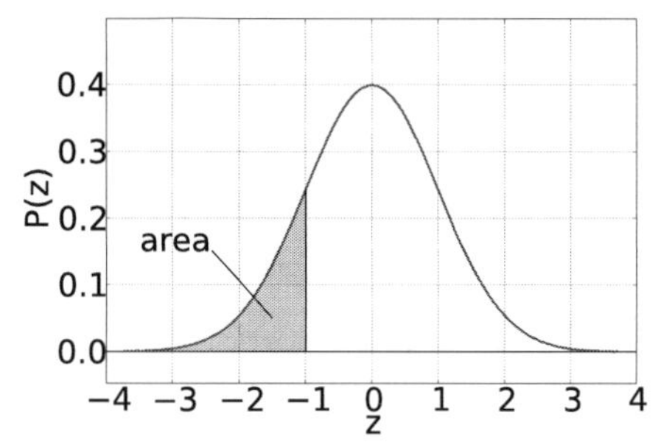

z	Area on Left	z	Area on Left	z	Area on Left	z	Area on Left	z	Area on Left
-2.20	0.0139	-1.90	0.0287	-1.60	0.0548	-1.30	0.0968	-1.00	0.1587
-2.19	0.0143	-1.89	0.0294	-1.59	0.0559	-1.29	0.0985	-0.99	0.1611
-2.18	0.0146	-1.88	0.0301	-1.58	0.0571	-1.28	0.1003	-0.98	0.1635
-2.17	0.0150	-1.87	0.0307	-1.57	0.0582	-1.27	0.1020	-0.97	0.1660
-2.16	0.0154	-1.86	0.0314	-1.56	0.0594	-1.26	0.1038	-0.96	0.1685
-2.15	0.0158	-1.85	0.0322	-1.55	0.0606	-1.25	0.1056	-0.95	0.1711
-2.14	0.0162	-1.84	0.0329	-1.54	0.0618	-1.24	0.1075	-0.94	0.1736
-2.13	0.0166	-1.83	0.0336	-1.53	0.0630	-1.23	0.1093	-0.93	0.1762
-2.12	0.0170	-1.82	0.0344	-1.52	0.0643	-1.22	0.1112	-0.92	0.1788
-2.11	0.0174	-1.81	0.0351	-1.51	0.0655	-1.21	0.1131	-0.91	0.1814
-2.10	0.0179	-1.80	0.0359	-1.50	0.0668	-1.20	0.1151	-0.90	0.1841
-2.09	0.0183	-1.79	0.0367	-1.49	0.0681	-1.19	0.1170	-0.89	0.1867
-2.08	0.0188	-1.78	0.0375	-1.48	0.0694	-1.18	0.1190	-0.88	0.1894
-2.07	0.0192	-1.77	0.0384	-1.47	0.0708	-1.17	0.1210	-0.87	0.1922
-2.06	0.0197	-1.76	0.0392	-1.46	0.0721	-1.16	0.1230	-0.86	0.1949
-2.05	0.0202	-1.75	0.0401	-1.45	0.0735	-1.15	0.1251	-0.85	0.1977
-2.04	0.0207	-1.74	0.0409	-1.44	0.0749	-1.14	0.1271	-0.84	0.2005
-2.03	0.0212	-1.73	0.0418	-1.43	0.0764	-1.13	0.1292	-0.83	0.2033
-2.02	0.0217	-1.72	0.0427	-1.42	0.0778	-1.12	0.1314	-0.82	0.2061
-2.01	0.0222	-1.71	0.0436	-1.41	0.0793	-1.11	0.1335	-0.81	0.2090
-2.00	0.0228	-1.70	0.0446	-1.40	0.0808	-1.10	0.1357	-0.80	0.2119
-1.99	0.0233	-1.69	0.0455	-1.39	0.0823	-1.09	0.1379	-0.79	0.2148
-1.98	0.0239	-1.68	0.0465	-1.38	0.0838	-1.08	0.1401	-0.78	0.2177
-1.97	0.0244	-1.67	0.0475	-1.37	0.0853	-1.07	0.1423	-0.77	0.2206
-1.96	0.0250	-1.66	0.0485	-1.36	0.0869	-1.06	0.1446	-0.76	0.2236
-1.95	0.0256	-1.65	0.0495	-1.35	0.0885	-1.05	0.1469	-0.75	0.2266
-1.94	0.0262	-1.64	0.0505	-1.34	0.0901	-1.04	0.1492	-0.74	0.2296
-1.93	0.0268	-1.63	0.0516	-1.33	0.0918	-1.03	0.1515	-0.73	0.2327
-1.92	0.0274	-1.62	0.0526	-1.32	0.0934	-1.02	0.1539	-0.72	0.2358
-1.91	0.0281	-1.61	0.0537	-1.31	0.0951	-1.01	0.1562	-0.71	0.2389
-1.90	0.0287	-1.60	0.0548	-1.30	0.0968	-1.00	0.1587	-0.70	0.2420

Cumulative Normal Distribution (cont.)

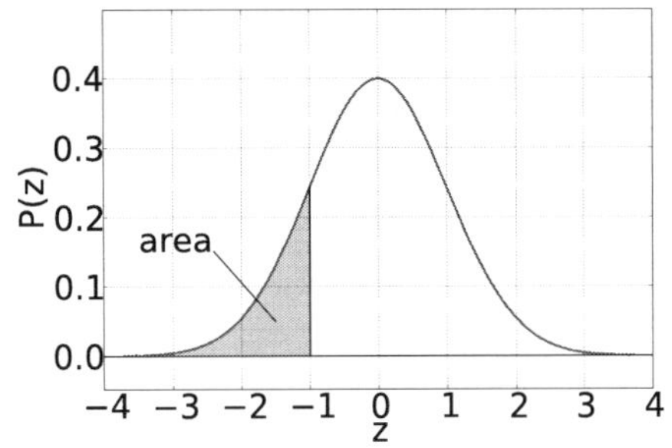

z	Area on Left	z	Area on Left	z	Area on Left	z	Area on Left	z	Area on Left
-0.70	0.2420	-0.40	0.3446	-0.10	0.4602	0.20	0.5793	0.50	0.6915
-0.69	0.2451	-0.39	0.3483	-0.09	0.4641	0.21	0.5832	0.51	0.6950
-0.68	0.2483	-0.38	0.3520	-0.08	0.4681	0.22	0.5871	0.52	0.6985
-0.67	0.2514	-0.37	0.3557	-0.07	0.4721	0.23	0.5910	0.53	0.7019
-0.66	0.2546	-0.36	0.3594	-0.06	0.4761	0.24	0.5948	0.54	0.7054
-0.65	0.2578	-0.35	0.3632	-0.05	0.4801	0.25	0.5987	0.55	0.7088
-0.64	0.2611	-0.34	0.3669	-0.04	0.4840	0.26	0.6026	0.56	0.7123
-0.63	0.2643	-0.33	0.3707	-0.03	0.4880	0.27	0.6064	0.57	0.7157
-0.62	0.2676	-0.32	0.3745	-0.02	0.4920	0.28	0.6103	0.58	0.7190
-0.61	0.2709	-0.31	0.3783	-0.01	0.4960	0.29	0.6141	0.59	0.7224
-0.60	0.2743	-0.30	0.3821	0.00	0.5000	0.30	0.6179	0.60	0.7257
-0.59	0.2776	-0.29	0.3859	0.01	0.5040	0.31	0.6217	0.61	0.7291
-0.58	0.2810	-0.28	0.3897	0.02	0.5080	0.32	0.6255	0.62	0.7324
-0.57	0.2843	-0.27	0.3936	0.03	0.5120	0.33	0.6293	0.63	0.7357
-0.56	0.2877	-0.26	0.3974	0.04	0.5160	0.34	0.6331	0.64	0.7389
-0.55	0.2912	-0.25	0.4013	0.05	0.5199	0.35	0.6368	0.65	0.7422
-0.54	0.2946	-0.24	0.4052	0.06	0.5239	0.36	0.6406	0.66	0.7454
-0.53	0.2981	-0.23	0.4090	0.07	0.5279	0.37	0.6443	0.67	0.7486
-0.52	0.3015	-0.22	0.4129	0.08	0.5319	0.38	0.6480	0.68	0.7517
-0.51	0.3050	-0.21	0.4168	0.09	0.5359	0.39	0.6517	0.69	0.7549
-0.50	0.3085	-0.20	0.4207	0.10	0.5398	0.40	0.6554	0.70	0.7580
-0.49	0.3121	-0.19	0.4247	0.11	0.5438	0.41	0.6591	0.71	0.7611
-0.48	0.3156	-0.18	0.4286	0.12	0.5478	0.42	0.6628	0.72	0.7642
-0.47	0.3192	-0.17	0.4325	0.13	0.5517	0.43	0.6664	0.73	0.7673
-0.46	0.3228	-0.16	0.4364	0.14	0.5557	0.44	0.6700	0.74	0.7704
-0.45	0.3264	-0.15	0.4404	0.15	0.5596	0.45	0.6736	0.75	0.7734
-0.44	0.3300	-0.14	0.4443	0.16	0.5636	0.46	0.6772	0.76	0.7764
-0.43	0.3336	-0.13	0.4483	0.17	0.5675	0.47	0.6808	0.77	0.7794
-0.42	0.3372	-0.12	0.4522	0.18	0.5714	0.48	0.6844	0.78	0.7823
-0.41	0.3409	-0.11	0.4562	0.19	0.5753	0.49	0.6879	0.79	0.7852
-0.40	0.3446	-0.10	0.4602	0.20	0.5793	0.50	0.6915	0.80	0.7881

Cumulative Normal Distribution (cont.)

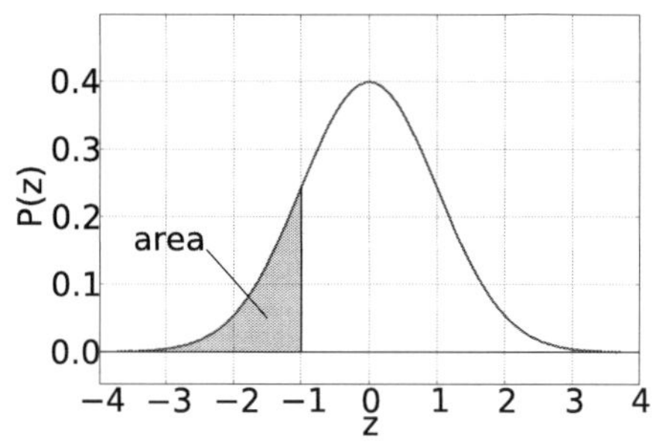

z	Area on Left	z	Area on Left	z	Area on Left	z	Area on Left	z	Area on Left
0.80	0.7881	1.10	0.8643	1.40	0.9192	1.70	0.9554	2.00	0.9772
0.81	0.7910	1.11	0.8665	1.41	0.9207	1.71	0.9564	2.01	0.9778
0.82	0.7939	1.12	0.8686	1.42	0.9222	1.72	0.9573	2.02	0.9783
0.83	0.7967	1.13	0.8708	1.43	0.9236	1.73	0.9582	2.03	0.9788
0.84	0.7995	1.14	0.8729	1.44	0.9251	1.74	0.9591	2.04	0.9793
0.85	0.8023	1.15	0.8749	1.45	0.9265	1.75	0.9599	2.05	0.9798
0.86	0.8051	1.16	0.8770	1.46	0.9279	1.76	0.9608	2.06	0.9803
0.87	0.8078	1.17	0.8790	1.47	0.9292	1.77	0.9616	2.07	0.9808
0.88	0.8106	1.18	0.8810	1.48	0.9306	1.78	0.9625	2.08	0.9812
0.89	0.8133	1.19	0.8830	1.49	0.9319	1.79	0.9633	2.09	0.9817
0.90	0.8159	1.20	0.8849	1.50	0.9332	1.80	0.9641	2.10	0.9821
0.91	0.8186	1.21	0.8869	1.51	0.9345	1.81	0.9649	2.11	0.9826
0.92	0.8212	1.22	0.8888	1.52	0.9357	1.82	0.9656	2.12	0.9830
0.93	0.8238	1.23	0.8907	1.53	0.9370	1.83	0.9664	2.13	0.9834
0.94	0.8264	1.24	0.8925	1.54	0.9382	1.84	0.9671	2.14	0.9838
0.95	0.8289	1.25	0.8944	1.55	0.9394	1.85	0.9678	2.15	0.9842
0.96	0.8315	1.26	0.8962	1.56	0.9406	1.86	0.9686	2.16	0.9846
0.97	0.8340	1.27	0.8980	1.57	0.9418	1.87	0.9693	2.17	0.9850
0.98	0.8365	1.28	0.8997	1.58	0.9429	1.88	0.9699	2.18	0.9854
0.99	0.8389	1.29	0.9015	1.59	0.9441	1.89	0.9706	2.19	0.9857
1.00	0.8413	1.30	0.9032	1.60	0.9452	1.90	0.9713	2.20	0.9861
1.01	0.8438	1.31	0.9049	1.61	0.9463	1.91	0.9719	2.21	0.9864
1.02	0.8461	1.32	0.9066	1.62	0.9474	1.92	0.9726	2.22	0.9868
1.03	0.8485	1.33	0.9082	1.63	0.9484	1.93	0.9732	2.23	0.9871
1.04	0.8508	1.34	0.9099	1.64	0.9495	1.94	0.9738	2.24	0.9875
1.05	0.8531	1.35	0.9115	1.65	0.9505	1.95	0.9744	2.25	0.9878
1.06	0.8554	1.36	0.9131	1.66	0.9515	1.96	0.9750	2.26	0.9881
1.07	0.8577	1.37	0.9147	1.67	0.9525	1.97	0.9756	2.27	0.9884
1.08	0.8599	1.38	0.9162	1.68	0.9535	1.98	0.9761	2.28	0.9887
1.09	0.8621	1.39	0.9177	1.69	0.9545	1.99	0.9767	2.29	0.9890
1.10	0.8643	1.40	0.9192	1.70	0.9554	2.00	0.9772	2.30	0.9893

Cumulative Normal Distribution (cont.)

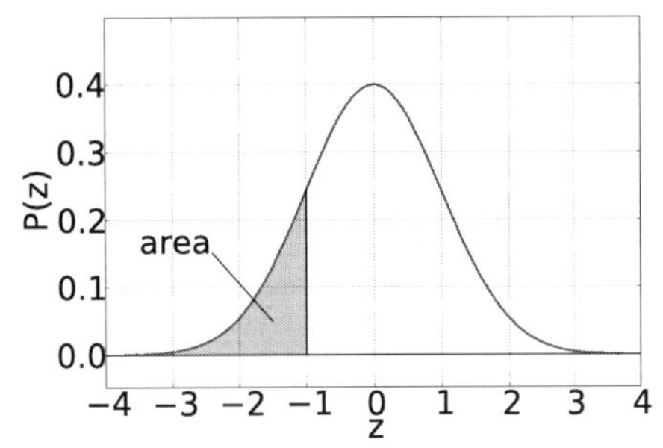

z	Area on Left	z	Area on Left	z	Area on Left	z	Area on Left	z	Area on Left
2.30	0.9893	2.60	0.9953	2.90	0.9981	3.20	0.9993	3.50	0.9998
2.31	0.9896	2.61	0.9955	2.91	0.9982	3.21	0.9993	3.51	0.9998
2.32	0.9898	2.62	0.9956	2.92	0.9982	3.22	0.9994	3.52	0.9998
2.33	0.9901	2.63	0.9957	2.93	0.9983	3.23	0.9994	3.53	0.9998
2.34	0.9904	2.64	0.9959	2.94	0.9984	3.24	0.9994	3.54	0.9998
2.35	0.9906	2.65	0.9960	2.95	0.9984	3.25	0.9994	3.55	0.9998
2.36	0.9909	2.66	0.9961	2.96	0.9985	3.26	0.9994	3.56	0.9998
2.37	0.9911	2.67	0.9962	2.97	0.9985	3.27	0.9995	3.57	0.9998
2.38	0.9913	2.68	0.9963	2.98	0.9986	3.28	0.9995	3.58	0.9998
2.39	0.9916	2.69	0.9964	2.99	0.9986	3.29	0.9995	3.59	0.9998
2.40	0.9918	2.70	0.9965	3.00	0.9987	3.30	0.9995	3.60	0.9998
2.41	0.9920	2.71	0.9966	3.01	0.9987	3.31	0.9995	3.61	0.9998
2.42	0.9922	2.72	0.9967	3.02	0.9987	3.32	0.9995	3.62	0.9999
2.43	0.9925	2.73	0.9968	3.03	0.9988	3.33	0.9996	3.63	0.9999
2.44	0.9927	2.74	0.9969	3.04	0.9988	3.34	0.9996	3.64	0.9999
2.45	0.9929	2.75	0.9970	3.05	0.9989	3.35	0.9996	3.65	0.9999
2.46	0.9931	2.76	0.9971	3.06	0.9989	3.36	0.9996	3.66	0.9999
2.47	0.9932	2.77	0.9972	3.07	0.9989	3.37	0.9996	3.67	0.9999
2.48	0.9934	2.78	0.9973	3.08	0.9990	3.38	0.9996	3.68	0.9999
2.49	0.9936	2.79	0.9974	3.09	0.9990	3.39	0.9997	3.69	0.9999
2.50	0.9938	2.80	0.9974	3.10	0.9990	3.40	0.9997	3.70	0.9999
2.51	0.9940	2.81	0.9975	3.11	0.9991	3.41	0.9997	3.71	0.9999
2.52	0.9941	2.82	0.9976	3.12	0.9991	3.42	0.9997	3.72	0.9999
2.53	0.9943	2.83	0.9977	3.13	0.9991	3.43	0.9997	3.73	0.9999
2.54	0.9945	2.84	0.9977	3.14	0.9992	3.44	0.9997	3.74	0.9999
2.55	0.9946	2.85	0.9978	3.15	0.9992	3.45	0.9997	3.75	0.9999
2.56	0.9948	2.86	0.9979	3.16	0.9992	3.46	0.9997	3.76	0.9999
2.57	0.9949	2.87	0.9979	3.17	0.9992	3.47	0.9997	3.77	0.9999
2.58	0.9951	2.88	0.9980	3.18	0.9993	3.48	0.9997	3.78	0.9999
2.59	0.9952	2.89	0.9981	3.19	0.9993	3.49	0.9998	3.79	0.9999
2.60	0.9953	2.90	0.9981	3.20	0.9993	3.50	0.9998	3.80	0.9999

Made in the USA
San Bernardino, CA
20 November 2014